SCARDED

D0042251

Chis Book

was donated to the

Richmond Public Library

By

The Friends of the
RICHMOND PUBLIC LIBARY

READING THE ROCKS

READING THE ROCKS

The Autobiography of the Earth

MARCIA BJORNERUD

Westview
PRESS

A Member of the Perseus Books Group

Copyright © 2005 by Marcia Bjornerud

Published by Westview Press
A Member of the Perseus Books Group

All rights reserved. Printed in the United States of America. No part of this book
may be reproduced in any manner whatsoever without written permission except
in the case of brief quotations embodied in critical articles and reviews.

Westview Press books are available at special discounts for bulk purchases in the
United States by corporations, institutions, and other organizations. For more information, please contact the Special Markets Department at the Perseus Books
Group, 11 Cambridge Center, Cambridge MA 02142, or call (617) 252-5298 or (800)
255-1514, or e-mail special.markets@perseusbooks.com.

Library of Congress Cataloging-in-Publication Data

Bjornerud, Marcia.
    Reading the rocks : the autobiography of the earth / Marcia Bjornerud.
      p. cm.
    Includes bibliographical references and index.
    ISBN 0-8133-4249-X (hardcover : alk. paper)
    1. Geology. I. Title.

QE31.B524 2005
551.7—dc22

                                                                2004022738

05 06 07 / 10 9 8 7 6 5 4 3 2

Contents

ACKNOWLEDGMENTS

First, I am deeply grateful to my editors—Holly Hodder, who picked up this odd pebble, and Ellen Garrison, Kay Mariea, and Patty Boyd, who polished it. I thank my colleague Jeff Clark (even though he likes rivers more than rocks) for keeping our department running smoothly when I have been away and for keeping me on solid ground the rest of the time.

I have been fortunate to have had many divinely earthy teachers and mentors who taught me to read and interpret rocks: Peter Hudleston (University of Minnesota) and Clement Chase (now at University of Arizona); Campbell Craddock, Robert Dott, and Gordon Medaris (University of Wisconsin); Yoshihide Ohta and Thore Winsnes (Norwegian Polar Institute); Håkon Austrheim and Inge Brynhi (University of Oslo); Hans Trettin (Geological Survey of Canada); and my late husband, Norman Grant. Each of them helped me to appreciate the nuances of the various rock dialects and shared with me the pleasure one can find in the company of stones.

Finally, my gratitude to my family is immeasurable (as all really important things are). They have never questioned the rocky path of my life. My three boys, Olav, Finn, and Karl—already veteran field geologists—are kindly indulgent of a mother who can't pass by a single stone without stopping to scrutinize it. My parents, James and Gloria, are our craton, the stable continental shield upon which everything else has been built. I dedicate this book to them.

Currently Accepted Geologic Timescale

(Time Intervals Not Proportional to Actual Length)

Era	Period	Epoch	Symbol	Began Million of Years Ago	Major Geological and Biological Events
Cenozoic	Quat-ernary	Holocene	Q	0.01 (i.e., 10,000 yrs)	Fossil fuel burning releases carbon dioxide stored for millions of years; Written history; agriculture
		Pleist-ocene		3	Ice Age
	Tertiary		T	65	Mammals diversify; Rockies, Andes, Alps, Himalaya form; Brief reign of giant birds
Mesozoic	Cretaceous		K	140	Mass extinction: Dinosaurs and others; Flowering plants and insects coevolve
	Jurassic		J	200	Dawn of age of reptiles
	Triassic		TR	250	Mass extinction; Modern Atlantic begins to open; Earth still reeling from oxygen crisis
Paleozoic	Permian		P	290	Permian oxygen crisis: Largest mass extinction in Earth's history; Ice age
	Carboniferous		C	355	Appalachians form; Pangaea assembled; Period of global warmth; lush forests; Thick coal beds in N. America, Europe
	Devonian		D	420	Modern fish evolve; Mass extinction
	Silurian		S	440	First ecosystems on land; First coral reefs, including Niagaran
	Ordovician		O	508	Mass extinction
	Cambrian		Є	545	Cambrian explosion as recorded in Burgess Shale and other fossil beds; Small shelly fauna
Precambrian	Proterozoic		pЄ	570	Ediacara disappear; predation emerges; Enigmatic Ediacaran animals thrive
				750	Snowball Earth ultra ice-age; Supercontinent Rodinia assembled
				1,000	One-celled achritarchs dominate seas; Midcontinent Rift, Lake Superior region
				1,200	First multicellular life
				2,000	Copper-nickel deposits at Crandon, Wisconsin, form at volcanic vents
	Archean			2,200	First eukaryotic fossils; Stratospheric ozone layer forms; Oxygen released by photosynthesizing bacteria begins to accumulate in atmosphere; banded iron formations
				2,500	Modern plate tectonics begins
				3,000	Oldest rocks in Wisconsin; Youngest Moon rocks; Granite-greenstone belts form in poorly understood tectonic events
				3,500	First direct evidence of life on Earth
				4,000	Oldest surviving rocks form (exposed in West Greenland, northern Canada)
	Hadean			4,400	Oldest surviving crystals form (Australia); Oldest Moon rocks form (highlands)
				4,500	Moon is formed when primitive Earth is struck by another differentiating planet; Earth differentiates into mantle, core
				4,600	Earth and other rocky planets form; Chondritic meteorites form

PROLOGUE:
STONE CRAZY

Like the dead-seeming, cold rocks, I have memories within that came out of the material that went to make me. Time and place have had their say.

— ZORA NEALE HURSTON,
DUST TRACKS ON A ROAD

NO PLACE WITH NO PAST

I AM A GEOLOGIST. I support my obsession with **rocks** by teaching at a small liberal arts college in that vast region of the United States called the Midwest.[1] Like many such colleges, this one was founded in the nineteenth century by idealistic philanthropists who hoped to bring enlightenment to the hinterlands. One hundred fifty years later, my colleagues and I remain faithful to this vision, but everyone agrees that nothing of significance—that is, nothing worthy of scholarly scrutiny—has ever happened here in this beer-drinking burg on the north shore of Lake Winnebago in Wisconsin.

It is true that Harry Houdini and Joe McCarthy (there's an unlikely pair) both lived here. So did Edna Ferber (you know, *Showboat, Giant*). And we can claim past coaches and players for the Green Bay Packers, locally beatified as saints. But that's

1

about it. Whatever history has happened here seems shallow and slightly degenerate. Even the name of the lake has been co-opted, transmuted into a byword for the excesses of American consumer culture.

The phone book and the highway map do hint that there may be a richer story. There are nearly as many Xiongs as Joneses living on and around Lake Winnebago in places called Oshkosh, Poy Sippi, New London, New Holstein, Vandenbroek, Fond du Lac. In 1730, just a few miles from here, French forces with the assistance of some Menominee warriors massacred an entire village of Sauk people in retaliation for raids on fur traders along the Fox River. There were so many corpses that the site of the massacre became known as Butte de Mortes, the hill of the dead, a name that persists today (although the local pronunciation, "Bewdahmore," has stripped it of some of its gravity).

And on the northeast shore of the lake, up on the high shelf of the Niagaran escarpment, there are effigy mounds, about a thousand years old, in the shapes of snakes, turtles, panthers. Panthers? To the southeast, mammoth bones with butcher marks have been dug out of sticky lake clay. Mammoths? The clay around here is thirty feet thick in places, deposited under a vast lake that formed late in the Ice Age. If that lake once had a name, it was forgotten millenia ago, so geologists have retroactively christened it Glacial Lake Oshkosh, after an Algonquin chief whose name has also been sewn onto dungarees and overalls for a hundred years.

Modern Lake Winnebago, at about 140,000 acres, is a sizable lake and the home of one of North America's largest populations of sturgeon—huge, primeval fish that can live to be a century old. But Lake Winnebago is just a tiny remnant of Lake Oshkosh, which once covered an area ten times bigger, a giant puddle formed as the great ice sheets melted. The clayey sediment that settled quietly out of Lake Oshkosh blanketed an older landscape of rolling hills and valleys—a hidden terrain that has been probed by drillers punching

holes through the impervious clay in search of well water for exur-
ban homes. The intractable clay turned out to be a good choice for
the county landfill, and when the site was excavated, backhoes un-
covered a layer of spruce twigs and branches more than twenty feet
down—vestiges of an Ice Age forest. A buried forest? The lake clay
stops abruptly over at the escarpment with the effigy mounds,
where the limestone cliffs suddenly jut up out of the flat landscape.
The rocks in the cliffs contain fossils of corals, sea lilies, and ancient
squids (squids?), records of a time when sea level was very high
and Wisconsin's latitude was very low.

West of town, on the opposite shore of the glacial lake, there is
a funny little outcrop of pink rock called **rhyolite**, the kind of
rock spewed out by active volcanoes like Mount St. Helens.[1] Volcanoes? The surface of the rhyolite was scoured by glacial ice, but
the rock itself is far older, ancient even to the squids that swam
by when the pink lavas formed an island in warm, sunlit Silurian
seas. And further west, along the Embarrass River, there are
coarse-grained granites as beautiful as those you'd see in the
Sierra Nevada. Their huge crystals suggest that they formed at
the roots of mountains (mountains?) deep underground. If you
turn north, you encounter more volcanic rocks, some glittering
with copper-nickel ores. Then you come to a ten-mile swath of
red iron formation that will make your compass crazy, formed at
a time when oxygen first became abundant in the air and iron
rusted out of the oceans. If you can keep your bearings and head
straight toward Lake Superior, you'll end up on basaltic lava
flows that put you in mind of Kilauea or Reykjavik.

Before the founding of our stalwart college, then, this small
patch of Wisconsin had seen influxes of immigrants, systematic
genocide, mammoth hunts, cataclysmic flooding, continental
glaciation, climate oscillations, ruthless erosion, marine inunda-
tion, luxuriant coral reefs, three generations of volcanoes, wholesale change in the chemistry of the atmosphere, and the rise of a

major mountain range. Other than that, nothing much has ever happened around here.

THE ACCIDENTAL DIARIST

The point, of course, is that this place, like every place on the planet, does have a past, but that some histories are quieter and less self-aggrandizing than others. Many—perhaps most—historical records are so vestigial as to be virtually invisible to those not attuned to subtlety. But the faintness of these records doesn't mean that the epochs they represent did not happen or do not "matter"; it is simply that no one had the luxury of chronicling these events in full. If the quiet histories of the hinterlands seem trivial or arbitrary, it is only because we haven't taken the time to understand how they unfolded—how landscape and cultures converged to produce mound builders, fur traders, a Houdini, and a Joe McCarthy in the same place, how climate and bedrock conspired to dictate the flow of glaciers in one age and the position of tropical reefs in another.

Like the place names on the highway map, which are a palimpsest record of human interaction with the land, rocks and landscapes are Earth's unsystematic chronicle of its past—its unintentional autobiography. Autobiographies are unavoidably subjective accounts of past events, blurred by imperfect memory, limited by myopia, and edited for aesthetic, egotistical, or legal reasons. To write an autobiography requires consciousness of self, and this by definition precludes the possibility of creating an objective and comprehensive chronicle. The one autobiography that has been recorded with no self-consciousness is Earth's own life story, written, very literally, in stone.

Unfortunately, stone has an undeserved reputation for being uncommunicative. The expressions *stone deaf, stone cold, stony silence,* and, simply, *stoned* reveal much about the relationship most people have to the rocks beneath their feet. But to a geologist, stones are richly illustrated texts, telling gothic tales of scorching heat, violent tempests, endurance, cataclysm, and reincarnation. Over more than 4 billion years, in beach sand, volcanic ash, granites, and garnet schists, the planet has unintentionally kept a rich and idiosyncratic journal of its past.[2]

In a time of ubiquitous marketing and image making, we may find comfort in the existence of such a ruthlessly neutral text. Our interpretation of it may be flawed or biased, but we can be sure that the writer has no agenda. If the narrative is amoral, apolitical, and indifferent, it is also ecumenical, egalitarian, and absolute. The story is larger than all of us, shaped by rules that antedate and supersede every economic, legal, and religious doctrine humans have ever created.

As literature, Earth's telling of its own story is an amateurish amalgam. The genre varies from breathless thriller to quotidian diary; the action ranges from microbial metabolisms to mountain building. Equal coverage is allotted to the fragrant beauty of life and the fetid details of death. Some events are covered at tedious length, others only fragmentally or obliquely, requiring the reader to piece together the plot and characters.

But since it is the story of our dwelling place, of how we came to inherit this commodious home from ancestors that lived and died during mainly good and some very bad times, it is arguably the one text that should be required reading for every earthling. We are dismantling the roof and altering the heating system in a stately old manse—our only source of shelter—before having bothered to understand the subtleties of its architecture. And we may be able to learn a thing or two about survival from a structure that has maintained itself in grand style for more than 50 million human lifetimes.

The rock record was not written as a collection of fables for our moral edification, but we would be foolish not to heed its wisdom. After millennia of experimentation, Earth and its myriad systems have learned to harness energies and balance polarities: mixing versus sorting, large versus small, cooperation versus competition, conservation versus innovation. If this balance had not been struck, life could not have persisted on the planet for nearly 4 billion years.[2]

Earth's equipoise is perhaps its most remarkable character trait. While Earth's sister planet, Venus, has simply grown hotter and hotter over time and Earth's brother, Mars, has slipped into a deep, cold sleep, our home planet has remained both awake and equable. The earliest entries in the rock record—a handful of Australian zircon crystals dated at 4.4 billion years—as well as all the subsequent volumes, clearly indicate that liquid water has been stable at Earth's surface from the very beginning of the planet's history.[3] Earth has had fevers and chills but has suffered no malady so extreme that its climatic immune system could not ultimately overcome it. This is because Earth has high- and low-tech strategies for mitigating crises, an ancient and astonishing system of checks and balances that involves the oceans, atmosphere, biosphere, and solid Earth.

The lessons we can draw from Earth's story are not merely metaphorical; rather, they are design archetypes that we should emulate in our economic and social systems if we wish to avoid irreparable instability. In writing about nature, we scientists diligently avoid the anthropomorphization of the phenomena under study, suppressing the natural tendency to see ourselves in everything we look at. But it should not be surprising that we recognize ourselves in the Earth, since we are its progeny. Our mistake is forgetting that we are simply the youngest children in a generations-old dynasty. Narcissistic fascination with our own short biographies blinds us to the far richer and deeper family saga. In reading the rock record, we may perhaps anthropomorphize the Earth a little if we also "geomorphize" ourselves, rediscovering the history of the Earth imprinted on us.

1

THE TAO OF EARTH

For water continually dropping will wear hard rocks hollow.

— PLUTARCH

EARTH IS A VERY PLEASANT PLANET and, according to her stone diaries, has been so for millions of millennia. It is easy to forget just how remarkable this condition is, in the same way that we tend to be unaware of good health until becoming sick. For at least four billion years, through meteorite impacts, climate changes, and continental reorganizations, liquid water has remained stable at Earth's surface, and life has thrived for nearly as long. Earth is a supersystem of countless smaller, interconnected systems involving rock, water, air, and life. These systems operate at spatial scales from microscopic to planetary, over time periods from seconds to millions of years. Anyone who has ever struggled with diabetes, depression, or debt knows how hard it is to achieve and sustain his or her own physiological, emotional, or financial equilibrium. The electrical blackout that hobbled the eastern United States and Canada in the summer of 2003 also illustrates how difficult it is to build stable systems of any complexity. How did a messy tangle of systems without a centralized control mechanism (e.g., a brain, a band director, or a board of trustees) maintain Earth's equipoise over time?

The fact is, we don't entirely know. But if this equilibrium had not prevailed, we would never have emerged to wonder about such things. Sentient life could only have emerged on a planet that has provided consistently and benevolently for its denizens. Little by little, we earthlings are beginning to understand some aspects of how this most peculiar planet works—what makes it so robust and resilient. Some of the key characteristics of Earth's systems are redundancy, recycling, and the capacity for self-correction. These attributes are common to the solid Earth, the oceans and atmosphere, and the biosphere.

THE DEPARTMENT OF REDUNDANCY DEPARTMENT: INERTIA AND SPARE PARTS

Earth systems are big, and the sheer number of players and the volume of matter lends stability to the system. At the planetary scale, Earth's size gives it greater thermal inertia than that enjoyed by the other rocky planets (a group that also includes Mercury, Venus, and Mars). Part of this thermal advantage arises from the geometric fact that large objects have lower surface-area-to-volume ratios than do smaller objects. Consider a cube whose sides have a length of L. The surface area of one side is $L \times L$, so the surface area of the entire cube is $6 \times L \times L$. The volume of the cube is, of course, $L \times L \times L$, so the surface-area-to-volume ratio is $(6 \times L \times L) / (L \times L \times L)$, or simply $6 / L$. This fraction gets smaller as L grows larger. Planets are not cubes, but the same principle holds for spheres. From a thermal perspective, more surface area means more heat loss. This is why mice have much higher metabolic rates than elephants do—it's simply harder to stay warm if you're small. As the largest of the rocky planets, Earth has lost less heat over time than have significantly smaller bodies like Mercury, Mars, and the Moon. One by one, according to their sizes, these worlds have gone tectonically dormant.

Earth has a second size-related thermal edge, one connected to a story geologists tell with schadenfreude: the tale of the physicist who fell from grace because he wouldn't listen to rocks. In the mid–nineteenth century, the eminent physicist Lord Kelvin (William Thomson) made a mathematically virtuosic effort to determine the age of the Earth. He based his theory on the assumption that the planet had begun as a molten sphere and had been steadily cooling ever since. Kelvin was the Einstein of his time, one of the high priests of the new science of **thermodynamics** (a temperature scale and a line of refrigerators are named in Kelvin's honor). It was quite natural that he would attempt to answer a question that was one of the scientific grails of the 1800s: How old is the Earth?

One implication of Darwin's new theory of evolution by natural selection was that immensely long periods had elapsed from the appearance of the first organisms to the present. Darwin himself fretted about this implication; contemporary estimates for the age of the Earth just didn't seem long enough. In *The Origin of Species*, he made a rough estimate by determining the volume of rock that had been removed by erosion to create the Weald, a broad valley in the south of England, and dividing that amount by an approximate rate of sediment transport. This "denudation of the Weald" calculation yielded a minimum age of about 300 million years.[1] Kelvin's calculations, however, which were far more sophisticated and mathematically unassailable, suggested that the Earth was only 20 to 40 million years old—about a tenth of Darwin's estimate (and less than one-hundredth of the presently accepted age). Darwin and other geologists of the time (Darwin considered himself a geologist) were stung by this apparently irrefutable number, which seemed so at odds with their understanding of the rock record. Kelvin was sneeringly dismissive of their protests, and for almost three decades, by which time Darwin had died, the physicist's refined estimate of 24 million years was accepted as the probable age of the Earth.

What Lord Kelvin didn't know was that Earth has not simply been cooling since its formation but has an internal heat source that keeps it warm. Earth's interior contains a great inventory of heat-producing radioactive elements—unstable isotopes like uranium-238 and potassium-40—which break down over geologic time and release heat energy that drives plate tectonics (the movement of the Earth's rigid outer layers as they slide over a more mobile interior). To Kelvin, the Earth looked young because it was still so hot. He can hardly be blamed for not incorporating radiogenic heat into his reckoning, since radioactivity had not been discovered at the time of his calculation (though the phenomenon was known for ten years before his death in 1907, and he chose not to consider its implications for Earth's age). What geologists find harder to forgive is his assertion that such a calculation superseded the conclusions reached by reading the Earth's account of its past.[2]

So Earth's size gives it a double thermal advantage: Not only does it have a comparatively small surface area through which to lose heat, but the planet also has a sizable store of radioactive elements to generate more. While Earth has experienced some net cooling over the past 4.5 billion years, its heat loss and heat production have remained more nearly in balance than on Mercury, Mars, and the Moon. These small worlds died young, but Earth still has a warm and mobile mantle that keeps its crustal plates dancing even at an advanced age.

In the biological realm too, the size and productivity of the biosphere is the secret to its success. Darwin's theory of the survival of the fittest was based on the Malthusian premise that population growth leads to a struggle for resources. Darwin knew that most plants and animals produce far more offspring than can survive, and he recognized that this superfecundity could be the driving force for evolution by natural selection. Individuals die but species persist, adapt, and thrive. The redundancy built into the system—the seemingly profligate seed production by a dandelion, the enormous number of eggs laid by a salmon—makes a biological community robust. Conversely, once the population of a species dwindles below some critical threshold, its reserves are too thin to allow it to keep pace with the processes that winnow individuals—disease, poaching, environmental change. When the survival of a species hangs on the lives of a few individuals, extinction is nearly inevitable.

EQUALS AND OPPOSITES

But this is not to say that bigger is better—only that big enough is essential. In fact, the stability of physical and biological systems requires countervailing forces that limit growth or define an optimum size. If every dandelion seed or salmon egg grew to maturity, the resources needed for future generations would be decimated. So scarcity acts both to encourage fertility and to limit population. Over time, organisms "calculate" their odds for survival and invest just enough in reproduction, which is energetically costly (resource intensive) for individuals, to assure a viable population in the next generation.

Unlike the size of a biological population, Earth's size is of course fixed, but it fortuitously represents an optimum for thermal stability. If Earth were significantly larger, heat production would exceed heat loss and the planet might never have developed the crispy outer crust necessary for a self-sustaining **plate tectonic** system.

The balance of forces governing other physical processes on Earth is equally remarkable. For example, mountains on Earth self-destruct at nearly the same rate that they are built. Two processes—erosion and gravitational collapse—act to keep peaks from becoming too lofty. As in a progressive taxation system, both of these natural processes exact the most from the highest mountains, enforcing a kind of topographic egalitarianism. Erosion, achieved mainly by water and glacial ice, is most effective where slopes are the steepest. Gravitational collapse occurs because the sheer weight of mountains exceeds the long-term strength of the rocks that form them, in the same way that a ball of Silly Putty gradually slumps into a pancake. The tallest mountains spread and flatten the fastest, creating an upper limit to topographic relief on Earth. The highest Himalayan peaks and Hawaiian volcanoes, both 5.5 miles in elevation (above sea level and the sea floor, respectively), probably represent this upper limit. Like the hero of a Greek tragedy, a mountain's defining attribute is also its downfall.

The remarkable thing is that it doesn't have to be so. We need only to look to Mars to glimpse an alternative reality. Mars once had both active volcanoes and liquid water. But because of its diminutive size (Mars has a diameter about six-tenths that of Earth), the red planet lost its interior heat to space faster than it could be regenerated. Its crust became a thick, cold rind, and its volcanoes became sluggish, ceasing to replenish the atmosphere with carbon dioxide and water vapor. In the absence of these greenhouse gases, the climate grew progressively colder. Today, any water that remains on Mars is thought to be locked into a permafrost layer just beneath the planet's dusty surface. With no erosion to tear them down and an exceptionally thick and rigid crust to hold them up, immense, sleeping volcanoes persist on Mars. In fact, little Mars has the largest volcano in the solar system—Olympus Mons (Mount Olympus), three times higher and broader than Earth's largest, Mauna Loa. Earth prunes volcanoes long before they get so big.

If Mars is a cold and dusty mausoleum recalling a more glorious past, Venus is a hellish maelstrom where things have never settled into a routine. About the same size as Earth but closer to the Sun, Venus has always been hot. Mere proximity to the Sun, however, is not enough to account for the broiling weather on our sister planet, where it's 860°F and cloudy every day. The carbon dioxide and sulfur dioxide exhaled by its still-active volcanoes have turned Venus into an asphyxiating hothouse, because no reciprocal process removes these gases from the atmosphere at rates commensurate with their production. Earth would have met the same fate had it not devised a way to extract greenhouse gases from the air and store them in sedimentary rocks. More than 99 percent of all the carbon in Earth's near-surface environment is stored in **carbonate rocks**—layers of limestone and dolomite—deposited from seawater. On Venus, carbon dioxide can leave the atmosphere only by dissipating slowly into space.

Although Venus and Mars are in some ways opposite extremes, they reached those extremes for the same reason: the lack of **feedback mechanisms** to modulate surface conditions. *Feedback* in the colloquial sense simply means a reaction or response. In scientific contexts, the word is used to describe a particular type of reciprocal interaction—a repeating process, or loop, in which the output (effect) of one stage becomes the input (cause) of the next. Feedback can take two very different forms—positive, or amplifying, and negative, or damping. **Positive feedback processes** are self-perpetuating and can have either good or bad results. If praising a child for good schoolwork causes the child to be even more studious, that is a positive feedback process with a positive result. On the other hand, criticizing a child for poor work may further diminish the youngster's motivation, an example of a positive feedback process with negative results. Winnie-the-Pooh observed that "the more it snows, the more it snows," your mother asserted that "virtue is its own reward," and all of us have noticed that the rich get richer. These vicious (or virtuous) circles and downward (or upward) spirals describe positive feedback phenomena. At its best, positive feedback leads to self-perpetuating progress (success, economic stability); at its worst, it produces self-amplifying volatility (nervous breakdown, hyperinflation, arms races).

Negative feedback, in contrast, acts to minimize change and dampen oscillations. Thermostats, shock absorbers, and mechanical governors are all devices designed to provide negative feedback. The system of checks and balances in the U.S. Constitution is really a collection of negative feedback mechanisms that give the government inertia and stability. Good health is maintained by innumerable negative feedback processes: Body temperature, blood sugar, immunity, and appetite, for example, are all controlled by processes that detect and correct for changes in the body. Physiologists use the word *homeostasis* to describe such biological maintenance of constant internal conditions. Failure of these homeostatic systems can cause debilitating or life-threatening conditions, such as hypothermia, diabetes, severe allergies, and eating disorders.

Earth's long-term planetary-scale stability can be attributed to analogous negative feedback mechanisms that operate across many spatial and temporal scales. A butterfly flapping its wings in China could, under very special circumstances (of positive feedback), affect the weather in Kansas, but on most days, a local gust of wind will quickly erase the butterfly's wake. Mountains meet their match in erosion and finite rock strength. Atmospheric carbon dioxide exhaled by volcanoes, a critical factor in global climate, is kept at a remarkably constant level by the concomitant precipitation of carbonate rocks (limestone) in the oceans. Among the rocky planets, Earth alone has developed these habits of self-control. Without them, conditions at Earth's surface would vary capriciously, and life might have found it difficult to evolve apace. Interestingly, however, life itself is an integral part of many of these control systems. For example, about half of the carbon dioxide that is now sequestered in limestone was extracted from the air over time by tiny marine organisms that incorporated the carbon into their shells. When these minute organisms died, their shells rained onto the seafloor. Over time, as sedimentation continued, the pressure of the overlying layers helped compact this shell material into limestone. Could the biosphere be the key to Earth's long-term equilibrium?

Going home to Mother Earth

Biospheric regulation of the planet is the central tenet of the controversial **Gaia hypothesis**—the proposal that the Earth can be viewed as a superorganism with the capacity to regulate its body chemistry and temperature. Although the concept of the Earth as a living being is perhaps as old as human culture, the formal scientific development of that idea began in the 1960s, when atmospheric chemist James Lovelock and philosopher Dian Hitchcock were hired by the National Aeronautics and Space Administration (NASA) as consultants for the *Viking* missions to Mars. Their job was to design an instrument or a method to detect any life-forms on the planet, which was then terra incognita. Lovelock and Hitchcock soon realized that such a device was unnecessary. Earth-based instruments had already shown that the thin atmosphere on Mars was nothing more than "volcano breath," which was strongly dominated by carbon dioxide, with no sign of modification by life processes.[3] The tenuous atmosphere of Mars, they argued, is chemically dead; the molecular mix is exactly what you would expect if the exhalations of ancient volcanoes had been allowed to evolve toward equilibrium and gradually disperse to space.

In their search for martian life, Lovelock and Hitchcock found themselves looking back at Earth. In striking contrast to that of Mars, Earth's atmosphere is not only completely different from volcanic emissions but also in a state of profound chemical disequilibrium, with easily oxidized compounds like methane and ammonia coexisting improbably with abundant oxygen. Such a metastable condition can be maintained only by continuous replenishment, in rather the same way that you can maintain balance on a bicycle only by continuously moving forward. Earth's peculiar mix of gases bears the signature of constant restocking by life processes: transpiration, respiration, digestion. Lovelock and Hitchcock emphasized that Earth's atmosphere is not an independent entity that is merely exploited by the biosphere. Instead, the biosphere continuously creates the atmosphere, which, like "a cat's fur, a bird's feathers, or the paper of a wasp's nest," is not living, but "a biological construction . . . an extension of a living system."[4] The simple but radical point is that life on Earth has not merely adapted to an inert environment but continuously modified that environment in ways that are in turn favorable for the perpetuation of the biosphere—an example of stable, positive feedback. At the suggestion of his friend and neighbor, the novelist William Golding, Lovelock named his idea the Gaia hypothesis after the Greek goddess of Earth.

The Gaia idea was hardly noticed by the scientific community when Lovelock first published it in a short paper in 1972 and in two longer ones written with microbiologist Lynn Margulis in 1974.[5] But in 1979, when Lovelock issued a book for a general readership, *Gaia: A New Look at Life on Earth* (Oxford University Press), scientific critics attacked his ideas with vehemence. The primary indictment, which rose largely from Lovelock's poetic language and not his scientific intent, was that the hypothesis was teleological; detractors claimed that his argument implied that the biosphere was designed with a purpose (planetary management) and administered by some sort of omniscient entity. Furthermore, critics charged, the idea was incompatible with evolution by Darwinian natural selection. It seemed impossible that individual organisms acting in their own self-interest could control Earth's environment on a global scale. Lovelock, with colleague Andrew Watson, countered these arguments with a computer model called **Daisyworld**.[6] The model showed how simple feedback processes could make it possible for individual organisms, quietly living and dying under the pressures of natural selection, to have the collective effect of maintaining global homeostasis.

Daisyworld is a hypothetical Earthlike planet, the same size as Earth and orbiting the same distance from a star similar to Earth's Sun. Like our Sun, this star has grown progressively brighter through time, radiating more and more heat. Yet the surface temperature on Daisyworld has remained nearly constant—and within the range tolerable to life—for most of the planet's history. This comfortable temperature is maintained because the biosphere on Daisyworld, which consists only of dark-, light-, and gray-colored daisies, has acted to moderate the temperature via a simple feedback mechanism. The daisies influence the surface temperature simply through their **albedo**, or reflectivity. Dark daisies absorb most of the sun's heat; light-colored daisies reflect much of it back to space. Gray daisies absorb about as much heat as they reflect. But how could the reflectivities of individual daisies affect the global temperature?

Early in the history of the planet, when the young sun was still relatively cool, global temperatures would be barely tolerable to life. At this time, dark daisies would be the fittest variety, because clusters of them would create local warm spots that would favor the growth of more daisies.

Soon, through positive feedback, the planet would be covered with dark daisies, whose collective effect would be to increase the global temperature well above what it would have been in the absence of life. If dark daisies were to continue to proliferate, the global temperature would exceed the tolerable limit, and natural selection would begin to favor somewhat lighter daisies, which would act to cool down the hot spots—an example of negative feedback.

At first, gray daisies would do better than very light ones because clusters of highly reflective light daisies wouldn't be able to keep temperatures warm enough for survival. Gradually, though, the sun's output would increase to the point at which light-colored daisies would become the fittest, because clusters of them would create cool oases that would favor the growth of more of their kind. If the local cooling were too extreme, natural selection would again favor some darker daisies.

Over time, then, the mix of daisy colors would be dictated by both positive and negative feedback between the daisies and the temperature conditions. In this way, individual daisies, without knowledge of, or concern for, the planet as a whole, would have acted to control the global environment. Ultimately, the heat produced by the sun would be so great that no type of daisy could moderate or tolerate the temperature, and all varieties would die out. But collectively, the daisies would have kept planetary temperatures within the tolerable range longer than in the absence of a biologically mediated feedback mechanism.

The Daisyworld model showed that biological feedback can, at least theoretically, regulate global conditions. But does this really happen on Earth? Testing the Gaia hypothesis has proven difficult. Unlike Daisyworld, Earth has millions of species and thousands of distinct ecosystems, which interact with their environments in myriad ways. Even when we can quantify the effects of biological activity on geologic processes, showing causality is difficult because of the disparity between human and planetary time scales. For example, although we know that biologically mediated storage of carbon dioxide in limestone is a major part of the long-term carbon cycle, we aren't really sure whether this represents a negative feedback process. That is, if carbon dioxide levels in the atmosphere increase significantly, will biological precipitation of carbonate minerals increase commensurately, thereby keeping atmospheric concentrations in check? Such questions have obvious relevance outside scientific circles. Meanwhile, the Gaia idea remains a source of lively debate within the sciences. Interestingly, biologists have been highly skeptical of the concept of a living planet, while geoscientists, who have traditionally studied nonliving things, are more receptive to it.

Even skeptics of the Gaia idea, however, acknowledge that it has inspired productive new lines of inquiry: **biogeochemistry**, **geomicrobiology**, and **geophysiology**. These highly interdisciplinary fields seek to understand the role of the biosphere in the flux of dozens of chemical elements and compounds through the Earth's atmosphere, hydrosphere, and lithosphere.

EVERYTHING OLD IS NEW AGAIN

Carbon, water, sulfur, phosphorus, and nitrogen are in constant motion at and near the Earth's surface, reincarnated again and again as minerals in rocks, gases in the atmosphere, ions in the ocean, schools of fish, leaves on trees. Each year, for example, even in the absence of human activities, about 440 million tons of carbon are transferred from one form to another, with about 45 percent of this carbon "remanufactured" and shipped by biological processes. Similarly, 5.8 billion tons of nitrogen and 740 billion kilograms of phosphorus change hands in a year. Organisms are involved in 87 percent of the nitrogen trades and more than 99 percent of the phosphorus transactions. Recycling is ubiquitous and obligatory on Earth. Even the Earth's **crust**, when it is old, is returned to the factory by the process of **subduction**, in which cold and dense rock sinks back into Earth's **mantle**. There is no natural equivalent of a landfill. Nothing is unusable waste, and nothing will last forever, at least not in any particular form. Matter resides temporarily in various lodging places, then moves on in new guises.

Each place, or **reservoir**, within a particular system (water, carbon, etc.) has a characteristic **residence time**–the typical duration of stay by its tenants. Residence times vary hugely even within a given biogeochemical system. A carbon atom may spend only a second in a human lung, then stay tens of millions of years in a layer of limestone. Water taken into the atmosphere by evaporation generally checks out again after just a week, whereas water in the deep polar oceans is content to abide there for millennia. Ocean crust can expect to reside at the surface for 150 million years or so before returning to the forge. Eventually, though, everything passes through the system. Earth's surface and subsurface—like our own skin and organs—are in a constant state of renovation, the overall architecture preserved even as the constituent parts are incrementally replaced. Nothing is permanent, and yet *because* of this, everything is eternal.

This is a very different kind of immortality than one finds on Mars, whose ancient volcanoes are eternal but apparently dead. The contrast is like the difference between a house that is lived in and one that has become a historical site. Inevitably, living in a house causes wear and carries unavoidable risks from which the furnishings of the museum are exempt. China breaks, bathtubs overflow, cooking fires blacken ceilings. Most of the time, these small disasters are easily remediated and become part of family lore. Earth has experienced countless domestic mishaps over time and quickly set things right again. Very occasionally, however, things go simultaneously awry in many systems at once, and the structural integrity of the entire edifice is threatened. If the damage is serious enough, rebuilding may be necessary. Not counting the havoc wreaked by rogue **meteorites** (which are just as likely to hit dead"museum"planets as living ones), the Earth has had at least two near-death experiences—times when biogeochemical cycles, ocean circulation, and climate became so deranged that the normal feedback mechanisms could not repair them. One was the late Precambrian **Snowball Earth** interval approximately 750 to 600 million years ago, an ultra–ice age when the oceans may have frozen over. The other was the Permian–Triassic oxygen crisis, which led to the most severe mass extinction event in Earth's history and in which some 90 percent of all species died out in less than a million years (more about both debacles in Chapter 4). Eventually the earth did recover from these nightmarish episodes (or you wouldn't be reading this), but both were followed by profound reorganizations of the biosphere.

THE EARTH FUGUE

In a sense, the Earth system owes its stability to being a big, endlessly repeating bundle of contradictions. The physical size of the Earth has allowed enough time for the emergence of an elegant, self-choreographed dance in which everything moves in circles and every tendency has its counterpoint. Each entity or action gains energy from an opposition: the individual from the collective, competition from cooperation, innovation from conservation, mixing from sorting. Over geologic time, the specific rhythms and idioms change slowly, but the essential rules of the dance remain the same: interaction, restraint, reincarnation.

These rhythms of nature shape all the organisms that evolved under their influence. It is folly to think that we can sit out the dance or make our own rules. It's rather like skipping school; we only hurt ourselves in the long run. Our bones evolved in the constant presence of gravity, and without this force to challenge them, they lose their strength. In the same way, without the force field of scarcity, a constant in our evolutionary past, we lose something of our full potential. Once we have enough to survive, we crave limits—the discipline of the dance. We understand deep in our bones that unchecked consumption and unchallenged political power are violations of ancient earth-law. The only uncertainty is what the penalties are.

One consequence is that the once vast Earth has been shrinking, at least in a relative sense. The magnitude of human actions on the Earth now matches those of natural agents. We are changing the underlying beat of the global dance. We quite casually double or treble local rates of erosion. We have increased global phosphorus fluxes by 10 percent. Human contributions to atmospheric carbon dioxide (from the burning of fossil fuels, deforestation, and concrete manufacturing) amount to about 8 billion tons per year—more than sixteen times the total annual emissions from the earth's volcanoes and four times faster than the precipitation of carbonate rocks from seawater. The Earth orchestra might obligingly increase the tempo of the dance in response to our activities, but since there is no conductor, communicating this to all players and sections could take many human lifetimes. The Earth might also take the opportunity, as it has in the past, to experiment unpredictably with other cadences for a few millennia before settling into a new riff, which we may or may not find to our taste (if we are still around to try it).

The uncertainties are immense, but if we wish to preserve our social, political, and economic structures, which don't weather surprises well, we need to understand the range of possible outcomes. Fortunately, Earth has kept a good record of what has happened in the past when biogeochemical upheavals have occurred. To read it, we need to speak the language of rocks.

2

READING ROCKS: A PRIMER

There are plenty of ruined buildings in the world, but no ruined stones.

—HUGH MACDIARMID
(CHRISTOPHER MURRAY GRIEVE)
SCOTTISH POET, 1892–1978

MEETING ROCKS ON COMMON GROUND

FOR MOST PEOPLE, collecting pretty pebbles on a beach is an aesthetic pleasure, a calming idyll. The stones are smooth, shiny, and multicolored—a fiery red one here, a tiger-striped one there. They clack together pleasantly in one's pocket. But to those with any understanding of the language of rocks, a walk along a pebble beach can be a maddening experience: All those stones, like intermittent transmissions from crackling radio stations, declaim a few words about their past. A recurrent phrase or melody may be recognizable amid the noise, but in aggregate, the effect is cacophonous. So, rocks are best observed at home, in their native habitats in the outcrop, where we can hear them clearly.

An outcrop is a place where bedrock, the skeleton of the Earth, is exposed at the surface and not obscured by vegetation, soil, or loose sediment. In areas that were covered by ice during the last Ice Age (these include much of North America), bedrock is commonly buried beneath thick glacial deposits. Frustrated geologists disrespectfully describe this material as *overburden* (glaciologists slyly retaliate by calling bedrock *underburden*). Finding outcrops is easy in arid and mountainous areas, where naked rock lies basking in the sun. But in humid and topographically subtle areas (think Indiana), outcrops are elusive. Invariably, the few rocks that do expose themselves become veiled with lichen over time (the resourceful geologist learns to note that certain colors of lichen signify particular rock types—orange for basalt, green for granite). Stream beds may provide the only fresh, natural exposures of bedrock, and geologists therefore seek out quarries and road cuts, where excavators and explosives have opened new windows into the Earth's geologic archives.

Once you have found a rock in its natural habitat, it is essential to identify a lingua franca shared with these survivors from ancient and unfamiliar times. How can we presume to understand objects that formed unimaginably long ago, when the very geography of the Earth was different? The first and most fundamental tenet of geology is the principle of **uniformitarianism**—which, in sound bite form, is the idea that the present is the key to the past. That is, we can use our understanding of processes occurring on Earth today to interpret rocks, the records of times in the past. While the concept of uniformitarianism may seem almost too obvious to warrant discussion, the articulation of the principle in the late eighteenth century was an intellectual revolution of the first order. And the idea is subtler than it first appears; the overzealous application of uniformitarian logic has sometimes blinkered geologists' vision.

Although the idea of uniformitarianism probably occurred to any number of people prior to 1800 (including Leonardo da Vinci, whose notebooks include perceptive sketches and descriptions of many geologic phenomena), the person generally credited with giving modern form to the concept is James Hutton. This Scottish gentleman farmer and physician was lucky enough to be on the fringes of the gifted group sometimes called the Edinburgh Enlightenment; his circle included economist Adam Smith, philosopher David Hume, Erasmus Darwin (grandfather of Charles), and James Watt, inventor of the steam engine.[1] In most of the classic geology texts, Hutton is a hero, "Saint Geo," the intrepid slayer of the dragon of **catastrophism**, the name given posthumously to theories that invoked exceptional or biblical events to account for landscapes and rock formations. Nevertheless, Hutton was apparently neither motivated by, nor even particularly aware of, contemporary "adversaries." He also is cast as a champion of scientific logic over religious irrationality, when in fact his interest in things geological seems to have sprung from a deeply felt spirituality. Moreover, he was in the habit of reaching his conclusions first and seeking evidence to support them afterward.[2]

As a landowner in a wet climate, Hutton was aware of how much soil was lost to the sea by erosion each year, and as a religious man, he was troubled by the thought that God would allow the continents simply to be worn progressively away. He therefore began to seek evidence for the rejuvenation of the land and intuitively understood that such evidence could be found only in rocks. He recognized that the rocks exposed on the seaside cliffs of eastern Scotland were formed from sediment that had been derived from older continental rocks. And in this single insight, the Scottish farmer simultaneously articulated the central precept of geology and made a compelling argument for an Earth that was far older than the 6,000 years allotted to it by the Church. In his one great treatise, "The Theory of the Earth," published in 1788, he showed remarkable understanding of modern geologic principles:

> The ruins of an older world are visible in the present structure of our planet, and the strata which now compose our continents have been once beneath the sea, and were formed out of the waste of pre-existing continents. The same forces are still destroying, by chemical decomposition or mechanical violence, even the hardest rocks, and transporting these materials to the sea, where they are spread out, and form strata analogous to those of more ancient date.[3]

Hutton died in 1797, the birth year of Englishman Charles Lyell, who was to become one of the most evangelical uniformitarians. In his influential three-volume work, *Principles of Geology* (published between 1830 and 1833), Lyell argued not only that natural laws have been constant over time (the essence of Hutton's view), but that the geologic processes governed by these laws have always been constant in intensity and rate. The subtitle of *Principles* effectively conveys the content of the 1,400-page opus: *An attempt to explain the former changes of the Earth's surface by reference to causes now in operation.* Lyell assembled examples from all over the world to bolster his argument for an ever-changing but oddly static "Peter Pan" Earth in which the one thing that did not change was the rate and nature of the processes of change. Lyell believed so fervently in this vision of a steady-state, cyclical Earth that he tried to explain away the growing evidence for biological evolution (*Principles* was written more than two decades before Darwin's *Origin of Species*). To accept that organisms had changed over time would be to acknowledge that Earth was in some sense aging and that some planetary-scale rules had changed over time.[4] (Fossils had long been understood to be the preserved remains of dead organisms, but in the 1830s, not all natural scientists recognized that some remains represented extinct species. Later in life, Lyell accepted the reality of organic evolution and became a friend and mentor to Darwin.)

Although Lyell's dismissal of biological evolution soon became untenable, his credo of strict uniformitarianism, sometimes called **gradualism**, remained deeply rooted in the collective subconscious of geologists well into the late twentieth century. Thus, for almost 150 years, interpretations of the geologic record that invoked events of magnitudes or rapidity unobserved in human history were generally dismissed as unscientific. It was not until the 1980s, when clear documentation for a giant meteorite impact emerged as the primary cause of the dinosaur extinction event, that the restrictive Lyellian doctrine was finally lifted. We must of course wear uniformitarian spectacles to read the rock record, but we recognize now that these are a little rose-colored. Over geologic time, events of nightmarish magnitude have occurred (though, thankfully, rarely), but such events do not violate the uniformitarian principle, since they too are governed by unchanging natural laws. In other words, what is uniformitarian from a long-term, planetary perspective may be cataclysmic from the human point of view.

A ROCK BY ANY OTHER NAME

Now that we have some confidence that we can interpret even very old rocks, the next step is to get to know them on a first-name basis. Those unfamiliar with rocks tend to describe any with visible crystals as granite, all white or light gray rocks as marble, and everything else as either sandstone or slate. This is understandable. The naming business of taxonomy is tedious, and geologists are admittedly too fond of neo(geo)logisms and poly-prefixed words. Still, without some sort of language with which to describe what we observe, interpretation is impossible. In this sense, geologic terms are a metalanguage—a vocabulary for describing the grammar of rocks as a first step toward understanding what they signify.

Ideally, we could come up with a tidy classification scheme that was unambiguous and universally applicable—the equivalent of a Linnaean biological system of first and last names for all creatures great and small. But unlike organisms, which all spring from a single common ancestor, rocks have many different origins and attributes. As a consequence, many rock-naming systems have been introduced, and all of them are at least a little untidy. Happily, everyone agrees on the three big categories of rock: **igneous** (formed from the incandescent molten state), **sedimentary** (deposited at Earth's surface and derived—ground down or eroded—from preexisting rock), and **metamorphic** (modified in the solid state by heat, pressure, deformation, or some combination of these factors). Hairsplitters can of course point to category-defying exceptions: Should a rock formed of volcanic ash that was transported and deposited by water be considered igneous or sedimentary? At what point exactly does the burial of loose sediment—a necessary stage in the formation of a sedimentary rock—actually become **metamorphism**? Part of becoming a geologist is to learn to tolerate fuzzy boundaries.

Within each of the three big rock groups are many rock types and even more rock names. The superfluity of names reflects the near impossibility of pure description. Virtually any classification system, if useful at all, is based on assumptions about which characteristics are the most significant or diagnostic. For example, to someone using rocks as building materials, "splittability" might be the single most important attribute. In fact, the rock names *shale, slate,* and *schist* all refer to the capacity of these rocks to be easily cleaved. *Shale* comes from the Old English root *scael,* which also evolved into *shell, scale, skull,* and the Norse toast *Skaal* (allegedly a reference to the Viking habit of drinking from skulls). *Slate* comes from the Old French *esclate,* meaning "to split or splinter"; the word *slat* has the same origin. *Schist,* a micaceous metamorphic rock, is a German import, but can be traced to a much older Indo-European word meaning "to cut or split." This root also lies behind *scissors, schism,* and *schizophrenia.* *Clay* describes the opposite sort of behavior and comes from an Indo-European root shared with *clod, clump, glob, glom,* and *glue.* But if you are interested in rocks as texts rather than as roofing tiles, the clumping-versus-splitting criterion is not particularly useful. Both sedimentary and metamorphic rocks can be strongly layered and splittable, but for completely different reasons. They may be found side by side in a drystone wall, but these rocks have very dissimilar stories to tell. So we need a system of classification that is based at least partly on our understanding of the genesis of rocks yet does not prevent us from thinking about alternative interpretations.

Geology as a discipline still bears the imprint of the nineteenth-century preoccupation with classification, an activity that provides a comforting sense of finitude and certainty in the face of nature's variability and ambiguities. The traditional subdisciplines within geology—mineralogy, petrology (the study of rocks), paleontology, stratigraphy (the study of layered sequences), geomorphology (the study of landforms)—essentially correspond with the names of the exhibit halls in geological museums of the late 1800s. Only very recently have these peephole subdisciplines begun to give way to more panoramic views of geologic phenomena through the lenses of new fields like biogeochemistry and paleoclimatology. But nomenclatures outlive the systems that spawned them, and over time the technical vocabulary of geology has become an idiosyncratic mélange of anachronisms, synonyms, and some genuinely useful terms.

Mark Twain recognized the richness of the geologic lexicon. In his *Autobiography* (1907), he describes a grizzled riverboat mate who salted his epithets with geologic terms:

> Being a chief mate, he was a prodigious and competent swearer, a thing which the office requires. But he had an auxiliary vocabulary which no other mate on the river possessed and it made him able to persuade indolent roustabouts more effectively than did the swearing of any other mate in the business, because while it was not profane it was of so mysterious and formidable and terrifying a nature that it sounded five or six times as profane as any language to be found on the fo'castle anywhere in the river service.
>
> [He] had no education beyond reading and something which so nearly resembled writing that it was reasonably well calculated to deceive. He read, and he read a great deal, and diligently, but his whole library consisted of a single book. It was Lyell's *Geology*, and he had stuck to it until all its grim and rugged scientific terminology was familiar in his mouth, though he hadn't the least idea of what the words meant. . . . All he wanted out of those great words was the energy they stirred up in his roustabouts. In times of extreme emergency he would let fly a volcanic irruption of the old regular orthodox profanity mixed up with and seasoned all through with imposing geological terms, then formally charge his roustabouts with being Old Silurian Invertebrates out of the Incandescent Anisodactylous Post-Pliocene Period and damn the whole gang in a body to perdition.[5]

There is an undeniable pleasure in the incantatory power of geological terms and the precise application of them to rocks one encounters ("Ah, what a lovely stylolitic micrite!"). But in this book, we will focus less on the language of geology and more on reading the rocks themselves, introducing rock names only when the vocabulary words facilitate our reading of Earth's autobiography, or when the names themselves shed light on the history of human discourse with stone.

GRAMMAR AND SYNTAX OF THE
THREE ROCK LANGUAGES

The three main types of rocks are like different literary genres. Just as you wouldn't look to a cookbook for information on military history, you wouldn't expect a sandstone to tell you much about the Earth's interior. Sedimentary rocks are the best reference works to consult if you are interested in past conditions at the surface of the Earth—for example, ancient climates, biological activity, or the distribution of water bodies. Igneous rocks chronicle the long-term chemical evolution of the Earth and provide glimpses into processes that occur at inaccessible depths. Metamorphic rocks, born in one setting (sedimentary or igneous) and transformed as they encounter new environments, are the travel writers of the rock world, chronicling their astounding journeys through the crust. So it is important to know which questions are appropriate to ask which rock, and how to phrase the query.

Sorting out sedimentary rocks

A sedimentary rock is a 100 percent recycled product made of materials derived from the weathering and erosion of preexisting rocks. If the material was deposited as physical particles by water (or wind or ice), the sediment is termed **clastic** (from the Greek for "broken pieces") or sometimes *detrital* (related to *detritus* and *detriment*, all from the Latin "to lessen"). **Sandstone** is a clastic sedimentary rock. Other sedimentary rocks, such as **limestone** and rock salt, were precipitated chemically—like rock candy—from water that had become supersaturated with dissolved atoms. These are termed **chemical** sedimentary rocks.

Whether clastic or chemical, sedimentary rocks are formed by hydrologic or atmospheric processes, so there are no sedimentary rocks on the waterless, airless Moon (other than a blanket of rock fragments from meteorite impacts). Recent high-resolution images of Mars show that ancient layered strata may be present there, consistent with other evidence that the planet once had liquid water at its surface. Perhaps Mars too once kept a sporadic diary of its activities.

Clastic sedimentary rocks are subdivided principally on the basis of the sizes of their constituent grains. The three main types, in ascending order of particle size, are **shale (mudstone)**, sandstone, and **conglomerate**. Using an alternative, alliterative nomenclature, we can call the types pelite, psammite, and psephite, from the Greek words for clay, sand, and pebbles (psephology, the study of elections, derives its name from the ancient use of pebbles for voting). Geologists even have precise quantitative definitions of particle sizes: Sand, in the technical sense, consists of grains between 0.003 and 0.08 inches ($1/16$ and 2 millimeters); progressively larger particles are classed as granules, pebbles, and cobbles. Anything larger than 10 inches (256 millimeters) is a boulder. Sedimentologists spend a lot of time disaggregating clastic sedimentary rocks and determining their grain sizes by passing the particles through ever-finer sieves. Appropriately, one of the leading sandstone experts of the last half century is Harvard's Raymond Siever.

Grain size is a sensible choice as the basis for classification because it reflects the agent of deposition (water, wind, or ice) and how fast that medium was moving. Water and wind efficiently sort sediments by size (actually weight, but this is equivalent to diameter if all the particles have the same composition and density), moving particles up to a certain size and leaving behind anything larger. A pebble conglomerate could only have been deposited in an environment where the current or wave action kept sand grains in suspension in the water. A mudstone tells us of a body of water so still that fine clay particles could quietly settle out of it. Conversely, an unsorted jumble of clay, sand, and cobbles fairly shouts "water did not organize me," and we can suspect that glacial ice, which does not discriminate on the basis of grain size, snowplowed the finest rock flour along with Volkswagen-sized boulders. Only a viscous mudflow rolling slowly down a slope could leave a deposit with a similar range of particle sizes. We would need to read the rock more closely to learn whether its origin was glacial or gravitational.

Grain size is only the first question to ask a clastic sediment. Through the *shape* of its grains, a sediment tells you how far away its source rocks were. Angular fragments indicate short travels with little time for rounding off sharp corners. Under the microscope, the surface texture of the grains also bears biographical clues: Windblown sand grains are pitted and "frosted" by countless collisions with other grains. Waterborne sands tend to have smoother skins.

Among the most revealing characteristics of a clastic sediment is its mineral composition. So-called hard-rock (igneous and metamorphic) geologists sometimes scoff that the study of sedimentary rocks is akin to learning about trees by examining sawdust. But on Earth, where erosion makes mountains ephemeral, their detritus alone may survive in the geologic record as the only information about a long-vanished landscape. Conglomerates in particular often contain chunks of older rock large enough to be read in their own right—fragmental texts from still more ancient times. Nineteenth-century geologists were able to reconstruct the main stages in the evolution of the Appalachian Mountains on the basis of the coarse clastic wedges—aprons of sediment that accumulated in low-lying areas—shed during each pulse of mountain building. The quartz-pebble *pudding stones* (conglomerates) of the Catskill Mountains in New York, for example, provide a detailed record of the dismantling of a great mountain belt.

Sandstones too may record information about an ancient continental terrain, but their grains tend to have been more winnowed by transport and weathering processes and must be interpreted through a statistical lens. Most sandstone consists mainly of the mineral quartz, not because quartz is the dominant mineral in crustal rocks (feldspars are) but because quartz is especially durable in the face of physical abrasion and chemical attack. Even more durable, but in far smaller quantities in sandstones, is a small group of igneous minerals whose presence can indicate the source, or provenance, of the sediment. These minerals include rutile (titanium oxide), tourmaline (a complex, boron-rich silicate), and zircon (zirconium silicate), all of which are derived from granitic igneous rocks, in which these minerals occur as minor constituents. Harvesting just a few grains of these needle-like minerals is truly like searching for a needle in a haystack, but the rewards can be great, particularly if you are combing ancient sandstones for zircon.

Here is why **zircon** offers such rich rewards: Zircon crystals are tightly sealed geologic time capsules, so tough that they can survive multiple cycles of weathering, transport, burial, erosion, and even high-temperature metamorphism (hot enough to melt lesser minerals) without losing the memory of their origins. In particular, a zircon crystal can yield a high-precision crystallization age based on the ratios of its radioactive uranium to uranium's daughter element, lead. This is why zircon grains have become grails for those interested in reconstructing Earth's most ancient landscapes. The **isotopic age** of a zircon crystal found in a sandstone does not tell you the age of the sandstone; rather, it reveals the age and nature of the rocks exposed at the surface at the time the sandstone was deposited. In fact, the very oldest discovered object native to Earth is a tiny zircon crystal picked from an ancient sandstone in western Australia. This venerable old crystal yields an astonishing uranium-lead age of 4.4 billion years, indicating that a stable, and possibly even granitic, crust had formed on Earth only 150 million years after the planet formed.[6]

If we stop focusing myopically on individual grains in a sedimentary rock and step back to see the pointillist picture the whole rock paints, the rock will reveal more about its origins. Is it layered? Are the layers horizontally continuous? Are they thick, thin, or variable? If variable, are the thickness variations random or repetitive? In clastic sedimentary rocks, layers usually represent distinct depositional events (e.g., the spring meltwater freshet in a river) or intervals of sediment accumulation punctuated by periods of nondeposition (e.g., a series of dry years when a shallow lake dried up).

A long-running philosophical debate in the field of **stratigraphy**—the study of sequences of sedimentary (stratified) rocks—is whether the sedimentary record represents mainly everyday happenings (e.g., waves and tides) or mainly extraordinary events (e.g., hurricanes and floods). The British stratigrapher Derek Ager advocated the latter, highly non-Lyellian interpretation, arguing that, like a soldier's life, the geologic record represents long periods of boredom interrupted by short periods of terror.[7] Ager pointed out that the day-to-day lapping of waves on the beach, for example, typically results in no sediment accumulation (just as much sand is washed out as washed in), whereas rare, large-magnitude events like hurricane Hugo leave piles of sand that are disproportionately represented in the geologic record. In the decades or centuries after a great storm, such deposits are reworked by ordinary waves and biological activity, creating what sedimentologists call a **palimpsest** layer. (The term is a particularly apt literary metaphor: A palimpsest manuscript is a parchment that was imperfectly erased and written over, from the Greek "to scrape again.") Ager called this crisis–quiescence cycle the Phenomenon of Quantum Sedimentation, and many strata that were once considered monotonous chronicles of steady accumulation are now recognized to bear the mark of violent storm deposition. Such rocks are given the rather Shakespearean designation **tempestite.**

But the sedimentary record isn't all Sturm und Drang. We see ample evidence of everyday events in many rocks in the form of sedimentary structures: Ripple marks recall a placidly flowing stream, a sinuous groove in graffiti left by an ancient arthropod scavenging food from the seafloor, honeycomb-like cracks in a shale record the soporific phenomenon of wet mud's drying. Some sedimentary rocks are even renowned and revered for methodical, consistent record keeping. These rocks, the reliable **rhythmites**, were deposited under conditions in which sediment supply was controlled by seasonal cycles, not sensational events of the disaster-movie genre.

Varves (from a Swedish term for "layers") are rhythmically banded deposits that accumulate in glacial lakes or fjords. In the summer, streams gurgle and flow, carrying sand and silt into the basin of standing water. In the winter, however, streams hibernate. Beneath the frozen surface of the water body, fine clays (the particles that give glacial lakes their otherworldly blue color) silently settle out. In this way, varves are rather like tree rings, with each centimeter-thick sand/clay layer pair representing the record of one year. In some Scandinavian lakes, continuous varve records go back as far 13,000 years. Buried with the sediments are other types of detritus: Pollen and spores chronicle changes in flora as the Ice Age ended; trace metals appear in medieval varves, documenting early lead mining. We geologists talk casually about millions and billions of years all the time, but there is still something eerie about being able to count backward from the present and put a calendar year on a geologic layer.

Tidal rhythmites found in rocks as old as 3.2 billion years record the tug of the Moon on the early Earth. In the best-preserved sediments, daily ebb and flow, twice-monthly neap and spring tides, and seasonal cycles can all be recognized. Astounding astronomical inferences can be made from these unassuming strata: The regularity of the ancient tidal deposits indicates a nearly circular lunar orbit, consistent with current theories of the Moon's formation by a giant collision between a Mars-sized planet and a still-forming Earth. The variation in grain sizes suggests fiercer tides and a Moon that was then much closer to Earth, again consistent with computer simulations of the Moon's violent origin. And perhaps most remarkably, the number of daily and monthly layers within a yearly packet makes it clear that Earth was spinning faster. A billion years ago, a day on Earth lasted only about 20.4 hours (though there would have been about 430 days in the year).[8] The tides themselves have been slowing the planet down, acting like giant hydraulic brakes on the spinning, solid Earth.

You just have to know the right rocks to ask.

Hard rock cachet

Unlike sedimentary rocks, which form in the familiar environment of the Earth's surface, igneous and metamorphic rocks come from the largely inaccessible, subterranean realm. "Hard rocks" have their own cryptic syntax, and reading them requires a bit more inference and perseverance. The direct application of uniformitarian logic is harder with these rocks, since we don't always know enough about present-day processes below the Earth's surface to understand the records of subsurface events in the past. But once we develop an ear for their idiom, these rocks can be translated.

The granite planet.   Classical gods still reign over igneous rocks, which are classified as either volcanic (emanating from Vulcan's forge onto Earth's surface) or plutonic (formed in the underground domain of Pluto). *Extrusive* and *intrusive* are equivalent, but less evocative terms. Crystal size is the key to determining whether an igneous rock cooled above or below the Earth's surface; it takes time for atoms in a magma to arrange themselves into the orderly rank-and-file lattices of minerals, and if the cooling rate is fast (as it is when lava is spewed from a volcano), crystals simply never have a chance to get organized. In the cases of extremely rapid quenching, no crystals develop at all. What forms is **obsidian,** or volcanic glass, an amorphous (noncrystalline) solid that is structurally still a liquid. **Pumice,** the spongelike volcanic rock that is so light it can float, is also glass—in this case, frozen froth from a gassy volcanic eruption (the word *pumice* shares an ancient Indo-European root with *foam* and *spume*). Pluto's rocks, in contrast, cooled slowly in his underground cellars and have coarse grains of distinct **minerals**—olivine, pyroxene, feldspar, mica, quartz—each formed over a particular range of temperatures.

The minerals in an igneous rock reveal much about the source of the melt from which it formed. All rocks in Earth's crust can be traced to an ultimate origin in the Earth's mantle, the rocky middle layer of the Earth that constitutes more than 80 percent of the planet's volume. But if you were to take a sample of typical mantle rock and melt it wholesale, it would produce a magnesium-rich, silicon-poor magma quite unlike any exuded by modern volcanoes. So there must be processes that selectively extract some elements from "raw" mantle, in a kind of planetary-scale distillation operation.

One such process is **fractional melting.** Imagine a queue of people waiting to buy tickets outside a ballpark on a hot summer day. As the sun rises higher in the sky, the heat grows unbearable to some. They relinquish their places in line and seek refuge in an ice cream parlor across the street, a small group of heat-intolerant people who have separated themselves from the heat-tolerant majority. Fractional melting works the same way. Consider a rock that is gradually heated to the point where only the least heat-tolerant components—typically those with more silicon and more large ions like calcium and sodium—begin to melt. Being less dense than the remaining, heat-resistant rock, the melt droplets may then rise to form a body of magma (a very hot ice cream parlor). In metallurgy, this same process, known as zone refining, is used to "sweat" the impurities out of industrial metals. In the Earth, fractional melting has created the oceanic crust, a layer of dark volcanic rock called basalt that covers two-thirds of the planet's surface.

Basalt consists mainly of the iron- and magnesium-rich mineral pyroxene, together with a calcium-rich feldspar called plagioclase. But igneous petrologists (*petra* is Greek for "rock," as in *saltpeter* or *petroleum,* "rock oil") often prefer their rocks pureed, describing the rocks' composition not in terms of mineral constituents but as percentages of their elemental oxide weights (Table 2.1). This is a bit like describing a loaf of bread as a particular combination of carbon, hydrogen, and oxygen rather than a mixture of flour, yeast, sugar, and water. But it does serve to highlight the geochemical lineages of the major igneous rocks.

The most pronounced trends in major element composition are the increasing content of silicon (Si), sodium (Na), and potassium

TABLE 2.1   Average Composition of Mantle Rock, Basalt, and Granite,
Expressed as Weight Percentages of Oxides of Major Elements

Element Oxide	Mantle Rock (Peridotite)	Basalt (Ocean Crust)	Granite (Continental Crust)
SiO_2	45.20	50.06	72.04
Al_2O_3	3.54	15.94	14.42
$FeO + Fe_2O_3$	8.48	11.40	2.90
MgO	37.48	6.98	0.71
CaO	3.08	9.70	1.82
Na_2O	0.57	2.94	3.69
K_2O	0.13	1.08	4.12
P_2O_5	trace	0.34	0.12

SOURCE: A. Ringwood, *Composition and Petrology of the Earth's Mantle* (New York: McGraw-Hill, 1975).

(K) and the sharply decreasing amount of magnesium (Mg) from the parental mantle **peridotite** (named for its dominant mineral, olive-green olivine, or peridot) to basalt to **granite**. Basalt can be produced in just one fractional-melting step from mantle material, and for this reason, basaltic crust is sometimes called secondary crust. The Moon's **lowlands**, or maria, as well as large regions on Mars and Venus are covered with basalts not dissimilar from those on Earth's ocean floor. This indicates that sometime in the past, partial melting occurred in the mantles of these other worlds, producing basaltic distillates.

But granite, with its pink potassic feldspar and pale quartz, is apparently unique to Earth. It is the rock type that constitutes much of the Earth's continental crust, yet is so different from peridotite that it cannot be produced in a single stage of fractional melting of mantle rock. Fractional melting of basalt, followed by fractional melting of the rock produced by that fractional melting, might yield a tiny amount of true granite. In other words, the very existence of granite in significant quantities on Earth is evidence of long-term planetary-scale refining and re-refining. It is as if, from an immense bowl of mainly green jelly beans, someone has systematically extracted the rare pink and white ones and spread them in a layer on the surface. Only Earth has produced such a "tertiary" crust so unrepresentative of the bulk composition of the planet's interior.

Consequently, one of the first questions to pose to an igneous rock is where it falls on the spectrum from primitive, or **mafic** (magnesium-rich, silicon-poor), to evolved, or **felsic** (magnesium-poor, silicon-rich). A mafic rock like basalt generally has tales to tell of life in the mantle, while for a felsic rock like granite, whose progenitors were themselves crustal, the mantle is a nearly forgotten ancestral homeland.

Basalts like those at oceanic hot-spot volcanoes like Kilauea are considered the least corrupted manuscripts available about the deep mantle. To read more from these rocks requires some initially counterintuitive techniques of deconstruction. First, since you know the rock is a basalt like any other (with a mineral composition of pyroxene and plagioclase and a major element composition as in Table 2.1), you discard all this information and look only at the trace elements the rock contains—the minute amounts of obscure elements like europium and ytterbium. The logic is that tiny differences in the concentrations of these trace elements are the fingerprints of processes at the source of the magma, just as subtle differences in the pronunciation of a few words can pinpoint the geographic region where an English speaker was raised.

The **rare-earth elements**, which range from lanthanum (atomic number 57) through lutetium (71) and occupy one of the two rows that hang awkwardly beneath the other elements in the periodic table, are favorite trace elements for analysis. These are present in rocks in vanishingly small, almost homeopathic, quantities as "impurities"—impostors standing in for major elements in the lattices of minerals. As the precision of geochemical instruments has improved, however, measuring concentrations of parts per billion has become routine in geological laboratories. This is like locating a few dozen particular people out of the entire human population on Earth.

Yet, geochemists analyze rare-earth element concentrations not so much to find what is in the rock but instead to reveal what was removed from it when it was still a melt! This Zen-like logic is a bit like inferring who has been eating from a box of assorted chocolates by studying what types are missing. Imagine that Peter has a known preference for nougats and caramels but eschews anything with nuts, while Crystal likes nuts and caramels but never touches nougats. Missing caramels may not be diagnostic, but nougats and nuts are. Similarly, different minerals have idiosyncratic preferences for different rare-earth elements. Garnet, a compact mineral formed only at high pressure, strongly favors the heavier rare earths, which have smaller ionic radii. Plagioclase accepts just europium, the only rare-earth element that has a valence charge of two and therefore "tastes" like calcium. Olivine avoids the whole lot if it can. By documenting what is missing from igneous rocks at Earth's surface, we can infer where they originated as melts and what must remain at those inaccessible depths. And by studying igneous rocks of different ages, we can understand how the chemistry of Earth's deep interior has changed through time.

You just have to know the right rocks to ask.

Metamorphic metaphors.   Metamorphic ("after-shaped") rocks are the polyglots of the petrological world, having resided in at least two distinct geologic settings in their lives. These rocks represent multiculturalism, not a melting pot: Metamorphism does not involve melting but rather recrystallization in the solid state, something like what happens to powdery new snow as it is buried and becomes crunchy. Consequently, the structures and compositions of metamorphic rocks are idiosyncratic hybrids of the environments they have inhabited, and this makes metamorphic rocks the richest of all geologic texts.

Metasedimentary rocks are typically the most readable type since they may preserve visible features like layering, ripple marks, and even **fossils**, which make it possible to identify the rock from the metasedimentary rock morphed—their **protolith** ("first rock"). This is like recognizing an aged friend you have not seen since childhood on the basis of a distinctive scar or the shape of an ear. But even if recrystallization and deformation have erased such features, the composition of a metamorphic rock records its original identity (even if outwardly changed, your friend remembers a long-ago summer with you at the beach). **Marble** is formed by heating limestone; both rock types are composed largely of the mineral calcite (calcium carbonate, or $CaCO_3$). The translucency of marble arises simply from the larger average size of the recrystallized grains. Slate, phyllite, and **schist** represent shales (mudstones) baked at progressively higher temperatures. Depending on the pressure and temperature conditions of metamorphism, the dull clays of a shale may become shiny mica, durable purple garnet, or sky-blue kyanite, all reconstituted forms of the aluminum and silicon present in the original clays.

Minerals like these, formed only under a relatively restricted range of physical conditions, are called **index minerals.** They are the imprints the rock received at various checkpoints in its life journey. By studying the index minerals, a geologist can trace the path of a particular rock from its origins, to its deepest burial, and back to the surface, where he or she happened to pick it up. A mineral like diamond, whose formation depends mainly on pressure, is a good **geobarometer,** providing a measure of the rock's depth at the time that the mineral formed. Other minerals crystallize only at specific temperatures and are used as **geothermometers.** These pressure- and temperature-sensitive minerals can survive as metastable components of their host rocks even as they travel toward the surface, just as a large pile of snow can persist for some time even when the temperature is above freezing. From a thermodynamic standpoint, however, diamonds are by no means forever. Alien to Earth's surface, they will slowly degrade to another, far more prosaic form of crystalline carbon: graphite, the soft mineral used for pencil "leads." Fortunately for jewelers and their customers, this degradation takes geologic intervals of time.

Index minerals are the key to recognizing the tectonic setting in which the rock was metamorphosed. As one burrows into the continental crust of the Earth, the temperature increases steadily at a rate of about 55°F per mile (20°C per kilometer). This change is observed directly in deep mines, where the temperatures in the lower levels can be sweltering. Some metamorphic rocks have mineral assemblages that accord with this **geothermal gradient**. That is, the temperatures the minerals record are just what would be expected at the pressures (depths) the rocks experienced. These rocks, which have matured in the conventional way, are said to have experienced ordinary, *burial metamorphism.*

But many other metamorphic rocks record peak pressure and temperature conditions that are not consistent with this typical geothermal gradient—the temperatures suggested by the rocks' composition are either too hot or too cold for the depths that these rocks reached. This means that the rocks underwent metamorphism under perturbed thermal conditions, the signature of magmatic or tectonic activity.

If a rock has index minerals that record anomalously high temperatures at low pressure—something like a child prodigy immersed too soon in the adult world—it must have been recrystallized near a heat source, probably a large body of subterranean magma. Such rocks are said to have undergone *contact metamorphism*. Conversely, if a rock contains high-pressure minerals (e.g., garnet, jade, or a rare diamond), yet never experienced commensurately high temperatures, then something refrigerated, or at least insulated, the rock while it was at great depth—rather like a naive adult who has led an unusually sheltered life. Rocks are exceptionally inefficient heat conductors, so it is possible for a block of rock, especially a large one, to ignore its thermal environment and remain cold even when surrounded by much hotter rock.

The one geologic setting in which such insulation might happen is a subduction zone, where cold ocean crust sinks back into the warm mantle, pulled by its own weight like a heavy blanket falling off a bed. The rate at which an ocean slab moves into the mantle (the downwelling part of a convection cycle) is many times faster than the rate at which it heats up (by conduction, which rocks do so badly). So ocean slabs can persist as anomalously cold sheets in the mantle for millions of years after they have been subducted and can even been 'seen' seismically, since earthquake waves passing through the Earth's interior travel a little faster through these cold zones.

Under poorly understood circumstances of planetary dyspepsia, rocks that went partway down a subduction zone sometimes find their way back to the Earth's surface. Such rocks are immediately recognizable by their signature high-pressure, low-temperature minerals. They are called **blueschists,** for the denim color of one of these, a sodium-rich mineral named glaucophane. Blueschists are very rare and have been granted far more than their share of pages in scientific journals, because they recall so unambiguously their journey into a subduction zone, sparing the rest of us the trip. Once again, you just need to know which rocks to ask.

Metamorphism related to subduction is certainly unique to Earth. With no tectonic recycling processes to move rocks from the surface to depth, the Moon, Mercury, Mars, and Venus presumably lack metamorphic rocks (unless you count shock-metamorphosed rocks traumatized during meteorite impacts). On Mars and Venus, extensive volcanism might have covered older rocks, causing these to undergo burial metamorphism, but the absence of efficient agents of erosion would keep such rocks mute at inaccessible depths and leave them unable to tell their stories at the surface.

MIND THE GAP: WHAT ROCKS DON'T TELL US

Collectively, sedimentary, igneous, and metamorphic rocks chronicle many different aspects of the Earth's evolution, in the same way that health records, school files, and family pictures provide distinct but complementary glimpses into a person's growth and development. But these records are necessarily incomplete and statistically nonrandom—they consistently exclude information about some events (e.g., everyday tasks) and over-represent others (e.g., exams, birthday parties). The geologic record too has its biases, and understanding what is missing is an important part of interpreting what is present.

(Re)moving mountains

James Hutton, the enlightened Scotsman who defined uniformitarianism, recognized the profundity of the silences in the rock record. In 1787 at Siccar Point, Berwickshire, near Scotland's border with England, Hutton made an observation that ranks as one of the great epiphanies of intellectual history.[9] At this windy, wave-sculpted promontory, Hutton noted two distinct sequences of layered rock: an upper sequence in which the strata were nearly horizontal (close to the orientation in which they were presumably deposited), and a lower sequence in which the layers were nearly vertical, standing on end like books on a library shelf. Between these two sequences was a nonplanar surface along which pebbles derived from the lower rocks were strewn in an irregular layer. Hutton later wrote that on seeing this surface, he "grew giddy from peering into the abyss of time," because he understood that it represented the countless centuries necessary to erode a mountain belt—whose roots were represented by the vertical layers—down to sea level (where the sediments of the overlying rocks were laid down). Although he did not know how much time this process would have taken, he knew that it probably represented more than that recorded by the rock layers themselves. In a clever effort to quantify the duration of this interval of erosion, Hutton used Hadrian's Wall, erected by the Romans in the second century to restrain marauding Celts, as a basis for estimating rock weathering rates. Except where whole blocks had tumbled from the wall, he saw that very little mass had been eroded from it in 1,700 years, which underscored his perception of the unfathomable depth of the "abyss of time." Assuming that the vertical rocks had also formed as submarine deposits before becoming folded into lofty mountains, Hutton saw evidence for an endlessly rejuvenating, infinitely old Earth with "no vestige of a beginning, no prospect of an end."

An erosional gap in the geologic record, like that described by Hutton at Siccar Point, is called an **unconformity**. Unconformities are ubiquitous and come in all magnitudes. Small ones, poetically called *hiatuses* and *lacunae*, represent sedimentary lapses of perhaps only millennia. At the other end of the spectrum is the

Great Unconformity of the lower Grand Canyon, where deep crustal metamorphic and igneous rocks are directly overlain by sediments deposited at the Earth's surface. The arrangement chronicles more than a billion years of erosional exhumation of these subsurface rocks prior to their reburial by the horizontal sedimentary strata for which the canyon is best known.

Unconformities record erosion, and erosion is a phenomenon that occurs principally on land, with the seafloor—especially the shallow continental shelves—being the ultimate repository for land-derived sediments. The sedimentary record therefore gives frustratingly short shrift to the land environments that are most familiar to us—mountains, plains, rivers—and disproportionately detailed coverage to shallow marine environments. Land or **terrestrial sediments** may be preserved only in nooks and crannies where erosion, a fastidious housekeeper, overlooks them (e.g., at the bottom of large lakes) or cannot keep up with their rate of accumulation (e.g., in the fan-shaped aprons of sediment at the foot of mountains). This underrepresentation of terrestrial environments in the sedimentary record also means that fossils of land-based biota are far less abundant than those of their marine contemporaries. Everything that is known about the celebrity dinosaur *Tyrannosaurus rex,* for example, comes from about fifteen complete specimens.

Old softies

Other creatures conspicuously excluded from most of the paleontological record are the scores of soft-bodied organisms—worms, jellyfish, and other wee beasts—that lack hard body parts like shells or skeletons (the soft parts of shelled or bony organisms are likewise rarely found). The benevolent oxygen that allows organisms to thrive in life consumes them ferociously in death. Nonmineral organic structures tend to decompose with little trace. As a result, most fossil beds do not preserve a cross section of the ecosystem that existed at a given place and time. The fossil record is an unrepentantly biased census taker. The problem of skewed sampling is so significant that an entire subdiscipline within paleontology is dedicated to it: **taphonomy** (from the Greek *taphos*, "burial" or "grave"). Taphonomists study modes and settings of fossilization in an effort to quantify the **preservation potential** of certain species and then apply statistical corrections to the fossil census to reconstruct the ancient population demographics more accurately.

The rarity of soft-bodied fossils is also the reason eighteenth-century geologists concluded that life on Earth had sprung rather suddenly into existence during the Cambrian period. Rocks older than this were notably devoid of shelly fossils, an observation that plagued Darwin and caused him to overemphasize the "Imperfection of the geologic record" (chapter 9 in *Origin of Species*). Ironically, Darwin's own statements about the incompleteness of the fossil record have been appropriated in some antievolution rhetoric today. Darwin earnestly believed that precursors to Cambrian organisms must have existed, and he would have been gratified to know that the field of Precambrian paleontology is now one of the most vibrant of all scientific disciplines. Powerful microscopes, together with analytical techniques that have allowed the identification of so-called chemical fossils—isotopic or molecular traces of organisms—show that Earth had been teeming with life for more than 2 billion years before the start of the Cambrian. The fossil record from this earlier era is simply much subtler because the Precambrian biosphere consisted mainly of very small, soft, anatomically simple life-forms. In fact, for most of this immensely long period, the only organisms on Earth were single-celled (yet incredibly diverse in metabolic strategies and lifestyles).

Against the odds, though, some soft organisms and body parts have found their way into the fossil record, preserved by the geologic equivalent of a hermetic seal: a lid of low-oxygen sediment. This rare preservation requires an unlikely combination of circumstances. The environment must have been stable enough to support a community of organisms—that is, typical sedimentation rates must have been low—but then a catastrophic sedimentation event had to inter the creatures without dismembering them in the process. These sediments, in turn, had to be buried by other layers and then evade erosion just long enough for humans wielding rock hammers to find them. Fortunately, over geologic time, odds of a million to one start looking pretty good, and the improbable happens. These jackpot beds are called **Lagerstätten**—German for "lode places"— because they are like striking gold.

Certainly the most famous fluke of this kind is the middle Cambrian Burgess Shale, found high on a mountainside near the continental divide in Yoho National Park, British Columbia, and named for a nearby peak, Mount Burgess. Created when a submarine mud slide swept a thriving community of organisms from shallow, sunny waters into the abyss, the shale preserves exquisite anatomical details of scores of species—the hard, the soft, and the partly digested—in proportions reflecting their natural populations. The stomach contents of predators allow dietary preferences to be documented. Together with familiar characters like **trilobites**, the shale contains bizarre-looking creatures (including *Anomalocaris* and *Hallucigenia*) that cannot readily be placed into known taxonomic categories. The presence of these curious fossils suggests that the middle Cambrian biosphere was far richer and more diverse than ordinary fossil beds even hint.

Most fossil-bearing layers are *death assemblages*—marine morgues containing biomineralized remains of creatures that had died sometime before they were incorporated into a sedimentary layer. In many beds with shelly fossils, for example, the shells tend to be uniform in size, not because the population was exceptionally homogeneous, but because the shells, like sand grains, were winnowed by wave action. The Burgess Shale, in contrast, is a *life assemblage*—the creatures were alive just moments before entombment. Like the ruins of Pompeii, the Burgess Shale provides a stunning glimpse of a moment from everyday life in the inaccessible past, unretouched by the normal geological editing processes. This rock unit, just a few yards thick, has arguably provided more information about the foundations of the modern biosphere (and has arguably been argued about more) than any other single stratum on Earth.[10]

Paradoxically, then, it takes an extraordinary event to enter the ordinary into the fossil record. But it does happen, and we are fascinated when we see the mundane, the scatological, and the pathetic in the fossil record: migration trackways left by ancient feet, fossilized dung (sanitized by the scientific name **coprolite**), unhatched eggs of extinct creatures.

Putting everything in order

These delightful cameo appearances in the fossil record, however, are not particularly useful in the larger endeavor of developing a worldwide chronology of events in the Earth's past. Since unconformities abound and no single place on Earth has a complete record of all of geologic time, we must find a way to determine how the fragmentary record of events at one place corresponds in time with the fragmentary records elsewhere. The problem is akin to finding hundreds of unnumbered pages from a jumbled manuscript and trying to put them in sequence. One needs to look first for chapter-level themes and eventually specific sentences that tie one page to the next.

In the early 1800s, William Smith, an English surveyor, canal digger, and self-trained geologist (identified somewhat roguishly in geological textbooks as William "Strata" Smith), realized that fossils provided this sort of global page-numbering system. At a time when the idea of evolution—not to mention the mechanism for it—was still controversial even among scientists, Smith documented a clear and consistent progression in the nature of the fossil organisms in the layered rocks of England. Although erosion had dissected these once-continuous marine strata, Smith was able to correlate, or trace, their isolated remnants by means of the characteristic fossils within them. Using this "principle of faunal succession," he methodically compiled his observations and in 1815 completed the first modern geologic map, a chart of England and Wales that is almost indistinguishable from those issued by the British Geological Survey today.[11] What is more important, though, he had begun to map geologic time, establishing survey markers in the previously uncharted "abyss of time" that Hutton had first glimpsed.

The key to Smith's method was the identification of **index fossils:** fossils of organisms that existed only for a comparatively brief period of geologic time. Species that survived for long intervals of time with little change—the most successful from an evolutionary standpoint—do not make good index fossils. Instead, the Edsels and eight-track tapes of the fossil record are the most useful, because they are diagnostic of well-defined geologic moments.

By 1880, careful **correlation** of fossils in sedimentary sequences from around the world had enabled scientists to create a global geologic timescale—a universal page numbering system—for at least the more recent chapters in the geologic record (Table 2.2). The chapters themselves were defined by watershed biological events. The beginning of the Paleozoic ("old life") era and its first subdivision, the Cambrian period, was identified by the sudden and abundant occurrence of shelly fossils (Darwin's bane). Everything before this was simply Pre-cambrian. The Paleozoic closed with a near-death experience for life on Earth: a severe **mass-extinction** event in which 90 percent of all marine species disappeared simultaneously from the fossil record. As the Mesozoic ("middle life") era dawned, this devastated biosphere gradually recovered, and by the Jurassic period, the terrible lizards reigned. Then another ruinous extinction punctuates the fossil record, marking the beginning of the Cenozoic ("recent life") era. Together, the Paleozoic, Mesozoic, and Cenozoic, chronicling both biological innovation and devastation, constitute the Phanerozoic ("visible life") eon.

The names of the geologic periods reflect the extent of the known geologic world in the mid–nineteeth century. *Cambrian* comes from the Roman name for Wales (*Cambria*), a latinization of the Welsh name, *Cymru. Ordovician* and *Silurian* also refer to Wales, taken from names of ancient tribes that lived in the Welsh mountains. The epochal subdivisions of the Ordovician and Silurian—Arenig, Tremadoc, Llanvirn, Llandeilo, Llandovery—are especially redolent of the vales of Wales. South across the Bristol Channel, the rocks of Devon became the type section for the Devonian period. The Carboniferous period is an exception to the generally geographic eponyms. It is simply a reference to the abundant organic matter—including the proverbial coals of Newcastle—preserved in strata from this time. The only Paleozoic period name that is not British in origin is the Permian, which comes from the mining town of Perm in the Ural region of Russia.

The Jurassic, thanks to Michael Crichton and Steven Spielberg, is certainly the most infamous division of geologic time outside academic circles. It takes its name from the idyllic alpine Jura region of Switzerland, where marine rocks of this age have been

TABLE 2.2   The Main Divisions of the Fossil-Based Geologic Timescale,
As Known by About 1880

Eon	Era	Period	Epoch	Notable Life Forms and Events
Phanerozoic	Cenozoic	Quaternary	Holocene Pleistocene	Written history Ice Age
		Tertiary	Pliocene Miocene Oligocene Eocene Paleocene	 Mammals proliferate
	Mesozoic	Cretaceous		Extinction of dinosaurs
		Jurassic		First birds
		Triassic		Reptiles proliferate
	Paleozoic	Permian		Mass extinction: 90% of existing marine species disappear
		Carboniferous		Much plant matter preserved in widespread coal swamps
		Devonian		Land plants begin to flourish Modern fish appear
		Silurian		First extensive coral reefs
		Ordovician		Proliferation of marine life
		Cambrian		Sudden appearance of diverse marine life with shells & skeletons
Precambrian	Proterozoic			Very sparse evidence of life
	Archean			

crumpled and uplifted as a result of the slow-motion collision of the African tectonic plate with Europe. The Triassic, a half-rhyme with Jurassic, is a homeless creation whose first syllable refers to the threefold subdivision of this period. The name, by coincidence, has echoes of *tragic* and *triage*, appropriate associations since the Triassic both began and ended with horrendous mass-extinction events. The Cretaceous, like the Carboniferous, is a reference to a widespread rock type of the age, in this case, chalk (in Latin, *creta*). The White Cliffs of Dover are Cretaceous chalks, but most of the epochs within the Cretaceous—Turonian, Coniacian, Santonian, Campanian—are named for places in France (Touraine, Cognac, Santes, Campania).

The two periods of the Cenozoic era represent rather embarrassing anachronisms in the geologic glossary: Tertiary and Quaternary are relics of a pre-Huttonian eighteenth-century system that divided rocks into four sequences: Primary or crystalline (granite, gneisses, etc.), Secondary (resistant, commonly folded, sedimentary strata), Tertiary (principally limestone and sandstones), and Quaternary (loose sediments). The continued use of the terms is akin to a modern doctor's employing the old nomenclature of the four humors in diagnosing a patient. For years, the International Commission on Stratigraphy, the vigilant arbiters of geologic timekeeping, has recommended that the Tertiary be retired. In the new version of the geologic timescale, the Cenozoic is apportioned into three periods: Paleogene (Paleocene through Oligocene epochs), the Neogene (Miocene and Pliocene), and the Quaternary (which remains as the lone survivor of the old system). But *Tertiary* is deeply embedded in the geolexicon, particularly as the *T* in the K–T (Cretaceous–Tertiary) boundary—the geologic moment when the dinosaurs went extinct. Geologists have long collective memories, and the old names will continue to crop up like stones in a plowed field, even though editors patiently clear them away year after year.

GETTING A DATE

The absolute ages and durations of the eras, periods, and epochs were not known until the twentieth century, when the discovery of radioactivity and the recognition of its utility as a geologic clock made quantitative age determinations possible. The principle is simple: The steady accumulation of radiogenic daughter atoms (e.g., lead-206, or ^{206}Pb in shorthand) relative to their radioactive parent (e.g., uranium-238, or ^{238}U) provides a proxy record of the time elapsed since a mineral crystallized (the time when the parent atoms were locked into the crystal lattice).

Imagine a very magnanimous parent who transfers half of his savings to his daughter each year on her birthday. Assuming that the father could continuously subdivide what remained of his wealth, he would never run out completely, although his reserve would grow very small. Conversely, the amount of money his daughter held would always increase, although the amount she received each year would grow smaller and smaller. At any time, an external auditor could determine the age of the girl—the number of years the parent had been giving money to her—by finding the ratio of the amount in the daughter's account to the amount in the father's account (excluding interest earned). This ratio would be independent of the absolute amount the parent had had at the beginning.

In nature, each species of radioactive parent **isotope** has its own characteristic payment schedule, or **half-life:** the time interval in which half of any collection of parent atoms will decay to the corresponding daughter product (in the preceding human analogy, the half-life was a year). Some radioactive isotopes (the ones humans should avoid) have half-lives of just decades, years, or days, but these have little geologic utility, since no matter how much is present at the beginning, vanishingly little parent material remains after about ten half-lives have elapsed. Other isotopes have half-lives of millions to billions of years, which constitute yardsticks long enough to measure geologic time spans.

Although the concept is simple, the practice of geochronology (literally, the science of Earth time) is fraught with difficulties both natural and technical. Although the first isotopic rock dates were obtained by heroic pioneers as early as 1915, calibrating the fossil-based, or biostratigraphic, geologic timescale has been an arduous process. And the task is by no means finished.

There is a bad joke that "geochronologists will date any old thing," but in reality, some old things are very hard to date. Only a handful of isotope systems and minerals are suitable for quantitative age determinations. One challenge is to find minerals that accept the parent isotope but not the daughter at the time of crystallization, or that at least provide some means of determining how much daughter material was there at the start. If some of the daughter isotope was present at the time of crystallization—that is, if the daughter in the monetary analogy had been given a large sum at birth—an auditor (geochronologist) might infer that she was older than she actually was. The way to get around this problem would be to somehow calculate and subtract the value of her account at the time of birth (crystallization). Also, the parent isotope must be abundant enough in Earth's crust to occur in measurable quantities in common rock-forming minerals. *Measurable,* however, is an operational term that has changed over time with improvements in analytical techniques. Exotic trace element parent-daughter pairs like rhenium-187 and osmium-187, present in such small concentrations that they were once beyond the limits of detection, are now part of the geochronologist's tool kit, making it possible to get dates from more old things.

A larger and more systemic difficulty in calibrating deep time is that isotopic dates give information about crystallization events, whereas the nineteenth-century geological timescale has biological events as its organizing principle. Pieced together via William Smith's inspired fossil correlation method, the original geologic timescale was based almost entirely on sedimentary rocks, since these are the bearers of fossils. Direct isotopic dating of sedimentary rocks, however, is not easy, since a date obtained from a constituent grain records not its time of deposition but rather the age of some older, usually igneous, rock from which the sediment was derived. Conversely, if a sedimentary rock is metamorphosed, new and datable crystals may grow in it, but these will be younger than the rock's depositional age.

The key to dating the sedimentological and paleontological events of the rock record, then, has been to find places where volcanic igneous rocks—either lava flows or, even better, regionally extensive ash layers—are interbedded with sedimentary strata. Gradually, the geologic timescale has been tuned and calibrated as dates have been obtained from volcanic crystals at sites where critical biostratigraphic boundary layers are exposed. As recently as 1993, 25 million years were lopped off the beginning Cambrian period (which jumped from 570 million years to 545 million years ago), when new uranium–lead dates were determined from zircon crystals in an ash layer along the Precambrian–Cambrian boundary in Siberia (see timescale on page 213).[12]

PEERING INTO THE PRIMORDIAL MISTS

The revised placement of the base of the Cambrian only added to the immensity of the Precambrian, that amorphous and largely unstratified part of the rock record that looked so opaque to Victorian geologists. Isotopic dating, particularly the uranium–lead dating of zircon crystals, has now allowed us to see the Precambrian in its proper proportions. By the 1940s, it was becoming clear that the Precambrian was many times longer than the Paleozoic, Mesozoic, and Cenozoic combined. The many named subdivisions of the Phanerozoic had caused a kind of temporal dysmorphia in the minds of geologists, exaggerating the length of time since the Cambrian and telescoping the time before it. Now recognized as eight-ninths of Earth's 4.5-billion-year history, the Precambrian is no longer viewed as the uneventful prologue to the inevitable rise of modern life, but as the main chapters in the sweeping tale of a planet where things might have turned out quite differently. In some ways, the Phanerozoic is the rather improbable epilogue to this much longer saga.

Although high-precision uranium–lead dating now allows the dating of events in the Precambrian with million-year resolution (which is astonishingly good for events that occurred more than a billion years ago), there is not yet a complete, standardized nomenclature for divisions of the Precambrian beyond the era level. The three big divisions, going backward in time are the following:

1. *Proterozoic:* runs from the base of the Cambrian (545 million years before present) back to 2.5 billion years ago (representing about 45 percent of the entire history of the Earth)
2. *Archean* ("ancient time"): from 2.5 billion back to about 4.0 billion years ago
3. *Hadean* (the hellish "unseen time"): defined with intentional flexibility as the period of Earth's history not recorded by any rocks surviving on Earth (although we do have Moon rocks of this age)

As older and older rocks have been discovered on Earth, the base of the Archean has been pushed back, and the Hadean has been shortened. For now, the august holders of the title for oldest preserved rocks on Earth are the 4.0-billion-year-old Acasta gneisses, exposed near Great Slave Lake in Canada's Northwest Territories.[13] As mentioned earlier in this chapter, ages greater than 4.0 billion years have been obtained for zircon crystals from western Australia, but the crystals are older than the metasedimentary strata that contain them and therefore don't really count as Hadean-age *rocks*.

Geologists disagree about the likelihood that any bona fide rocks older than 4 billion years still exist on Earth (or if the rocks do exist, whether they are anywhere that geologists could find them). According to the 1,500 pounds of lunar rock brought back by Apollo astronauts, the Hadean was not a pleasant time. A roiling magma ocean covered the Moon (and, we assume, likewise the Earth), solidifying slowly from the top down, only to have its primitive crust disrupted repeatedly by huge meteorites hurtling in at supersonic speeds after violent collisions with other space debris. The lunar maria, the vast, dark "seas" on the Moon first described by Galileo, are holes punched in this early crust. Their names—Tranquilitatis, Serenitatis, Fecunditatis, Nectaris—seem almost cruelly ironic in light of their violent origin and subsequent sterility.

Until and unless rocks of equivalent age are found on Earth, lunar names (e.g., Nectarian, Imbrian) have been adopted as subdivisions for the Hadean. But even the Moon lacks rocks from the very beginning, and so this elusive first chapter is therefore named the Cryptic. How, then, has the age of the Earth (and the Moon) been calculated to be 4.55 billion years—more than one hundred times older than Kelvin's estimate (but a little shy of Hutton's guess of infinity)?[14] The determination of this age is somewhat counterintuitive: Extraterrestrial rocks—meteorites—are the source of this isotopic age, which was determined by California Institute of Technology (Cal Tech) geochemist Clair Patterson in 1956. The logic is that these rocks formed at the same time that the other objects in the solar system did, but unlike any rocks on Earth or even the Moon, meteorites have remained unchanged since the beginning. (See also Chapter 4).

Apart from the largest subdivisions of the Precambrian (Archean and Proterozoic), no naming scheme for the oldest part of the earthly rock record has really taken root in the geologic literature. Several systems have been proposed, however, and the International Commission on Stratigraphy has issued its official recommendations. This seeming indifference to nomenclature is largely because isotopic dates eliminate the need for such names. The difference is rather like the contrast between British and North American addresses: The Geological Society of London, for example, receives its mail at Burlington House, Piccadilly, while the Geological Society of Canada has offices at 3303 33rd Street NW in Calgary. Although the first is rich with history and charming idiosyncrasy, the second is simply easier to find. For the same reason, the Precambrian will probably remain the domain of numbers, not names.

Translating the geologic record is a Herculean task that has taken thousands of interpreters more than two centuries and still is far from complete. Pages, if not entire chapters, of the Earth manuscript remain undiscovered, and passages from many known sections have not been fully deciphered. To those involved in the translation process, every pebble on the beach is a scrap of text from this immense and venerable tome, and it seems almost rude to ignore what these objects are saying. Once acquired, the habit of reading rocks is very hard to break.

3

THE GREAT AND THE SMALL

How deep is the ocean? How high is the sky?

—IRVING BERLIN

It's a small world after all.

—RICHARD M. SHERMAN AND
ROBERT B. SHERMAN

IN SCIENCE, to measure is to understand—an equation made plain in the dual meanings of the verb *to fathom*. The unfathomed is literally the murky unknown. Measurement is also the beginning of management; assaying, surveying, or weighing an entity confers a sense of control over it. Measurement would seem to be about accurate assessment of size or amount, but what measurement does, conceptually, is shrink things too immense to study or magnify things too minute to see so that they can be manipulated as numbers. Measurements allow us to draw maps and build models, manageable versions of systems too large or intricate for us to consider at their true scale. Making geologic measurements can be unexpectedly problematic, however, for reasons ranging from the practical difficulties of defining standardized units and building reliable instruments to the intrinsic immeasurability of some phenomena. In attempting to appraise the Earth, we are discovering, somewhat disturbingly, that our perceptions of magnitude are dependent on our measurement techniques. Like a set of Chinese boxes, Earth systems seem to grow as we look inside them. As we open each box, we are finding that the opened box somehow contains more than the one that enclosed it. Our traditional notions of size, rank, and hierarchy are being turned inside out. We have far to go before we completely fathom the Earth.

GEO-METRY: SIZING UP THE EARTH

To develop a theory for how any system works—be it a machine, an organization, or a planet—we must have some sense of its scope and proportions. How big is it? How many distinct components does it have, and how are these connected to each other? How fast does energy, matter, or information flow through it? The first step in answering these questions is simply to create practical metrics for describing essential characteristics.

Measures of length or distance were probably among the first to be invented, as suggested by the plethora of units in the Western tradition alone. The old measures are wonderfully organic and idiosyncratic. Many, logically, are based on the body as a ruler: for example, the foot, the hand, the cubit (the length of the forearm from the elbow to the tip of the middle finger), and the fathom (from the Old English *faethm,* "outstretched arms"). The yard, though at one point officially defined as the distance from King Henry VIII's nose to his outstretched fingertip, shares an origin with the words *girth* or *gird* and refers to the circumference of a person's waist (probably not Henry's ample one). Longer units refer to distances covered in motion: Furlongs are furrow lengths; the mile was defined by the Romans as one thousand pace (double-step) lengths.

In light of the ubiquitous use of the human body as a standard for measurement, surprisingly few ancient measurement systems used the decimal basis suggested by the ten fingers on human hands. Even more interesting, although the ancient Greek word *geometry* literally means "Earth measure," the first length unit to make direct reference to the size of the Earth itself is the meter. Originally defined in 1791 by the French Academy of Sciences, a meter was defined as one ten-millionth of a meridional quadrant of the Earth (the distance from one of the poles to the equator, along a line passing through Paris). Within a century, the emerging field of *geodesy* ("Earth division") had led to improved measurements of the figure of the Earth. The new measurements showed that the meter as defined by the academy was not actually one ten-millionth of the pole-to-equator distance. A physical standard—a bar of platinum and iridium called the Meter of the Archives—was then defined as the archetype of the meter. Since 1983, however, the meter has been defined in a way that disconnects it entirely from its Earthly origin: as the distance traveled by light in a vacuum in 1/299,792,458 of a second. This definition is operationally consistent with the yardstick used to measure cosmic distances: the light-year (about ten million billion meters). Ironically, the only scientific unit of length that remains Earth-based is the astronomical unit (AU), the average distance from the Earth to the Sun. The actual Earth-Sun distance not only varies over the course of a year because the Earth's orbit is elliptical but also changes over a 100,000-year cycle, during which the orbit's ellipticity grows and shrinks.

Earth's breadth and heights and depths—simple linear measurements that could be made by a remote observer—reveal much about the planet's long-term habits, in the same way that one can recognize an athlete or a couch potato at a glance. Earth is an oblate spheroid whose equatorial radius (3,961 miles, or 6,378 kilometers) is about 0.3 percent larger than its polar radius (3,948 miles or 6,357 kilometers). The planet's modest midriff bulge reflects the results of more than 4 billion years of daily rotation on the planet's deformable interior. Venus, which spins far more slowly—a day on Venus is somewhat longer than its year—has no such bulge (a cataclysmic collision with an asteroid early in the history of the solar system probably perturbed Venus's rotation, which is not only slow but backward).

Earth's inconsistent girth is large enough to cause bending stresses in the rigid tectonic plates as they move from high latitude to low latitude or vice versa, since the curvature of the Earth is not uniform. The planet's deviation from a perfectly spherical shape is also great enough to make the average pull of gravity smaller by about 0.5 percent at the equator than at the poles. That is, you actually weigh less at sea level in the tropics than you would in the arctic because, at the equator, you are slightly further from the center of the Earth. Most significantly, the gravitational interaction of the Sun and Moon with Earth's rotational bulge causes the Earth to precess, or wobble, like a spinning top over a cycle of 26,000 years, its rotation axis sweeping out a double cone shape in space. This slow roll has profound effects on the way solar energy is distributed around the Earth and is thought to be one of the major variables governing climate change. So even a seemingly trivial deviation from perfect roundness—26 miles on a planet that is nearly 8,000 miles across—has surprisingly large consequences.

Another noteworthy characteristic of Earth's figure is its unusual **hypsometry**—the distribution of elevations of the solid surface of the planet. This is usually depicted on a **histogram**, the kind of graph used to show the distribution of grades in a class or the numbers of people in different height intervals. On a planetary hypsometric plot, the horizontal axis is surface elevation relative to a datum (sea level for Earth), and the vertical axis is the percentage of the land surface that falls within a given range of elevations.

When elevation data for Venus and Mars are plotted in this way, the data define a single dromedary hump—a bell curve indicating one dominant average altitude flanked by higher and lower outliers. Earth's plot, in contrast, is Bactrian, that is, two-humped. This bimodal plot indicates that there are two distinct populations of elevations, one clustered at about 0.5 mile above sea level, and another around 2.5 miles below sea level, both with tails trailing off to either side. These outliers represent the highest highs (the Himalayas) and the lowest lows (ocean trenches) on the planet's surface. Without knowing anything else about the Earth, an observer would have to conclude that the outermost shell of the planet was composed of two completely distinct types of material with contrasting **density**, or buoyancy, relative to the substrate beneath, just as a low-density iceberg floats higher than a barge laden with iron ore. The crusts of Venus and Mars, conversely, would seem compositionally uniform and presumably basaltic.

While there are volcanic peaks and low-lying valleys on these sibling planets, these outlier elevations are just extreme end members along a continuum of elevations with one well-defined central value. Only Earth shows evidence of crustal differentiation; the two humps on its hypsometric plot represent the single-stage basaltic melts that form the ocean crust and the tectonically distilled granite of the continental crust. Earth's tectonic calisthenics have given it the planetary equivalent of good muscle definition—unusually pronounced topographic contrasts.

But there are so many other dimensions to a planet that is hydrologically, tectonically, thermally, and magnetically active. Pioneers in the study of particular geophysical variables sometimes earn the unique honor of becoming the namesakes for specialized units of measure. For example, Earth has about 1.4 billion cubic kilometers (0.34 billion cubic miles) of water, 97 percent of which is in the sea. This number alone, however, does not convey the dynamic nature of the oceans, where an immense amount of water is constantly in motion, carrying salt and heat around the globe. The flow rate for a major surface current like the Gulf Stream is typically 350 to 3,500 million cubic feet (10 to 100 million cubic meters) per second (for comparison, the lower Mississippi even at flood stage rarely exceeds 700,000 cubic feet per second, and Niagara Falls is a trickle at 100,000 cubic feet per second). To streamline their calculations and to honor one of the founders of their field, oceanographers have defined the *sverdrup* (Sv), a flow unit equal to 1 million cubic meters per second, for the great Norwegian scientist Harald Ulrik Sverdrup (1888–1957).

The flow of water through sediments and rocks in the subsurface is a much slower and more tortuous process. **Groundwater** flow was first described quantitatively by Henry Philibert Gaspard Darcy (1803–1858), a French civil engineer who designed a revolutionary gravity-driven water system for the city of Dijon. The mechanism provided water at the astonishing rate of 8 cubic meters (280 cubic feet) per minute. Darcy's greatest scientific contribution was to define a physical law that describes the rate at which a fluid will flow through a porous medium like sediment or fractured rock. One of the principal terms in that equation is the **permeability** of the rock or sediment—a measure of the connectedness of the open spaces within it. The scientific unit of permeability, used every day by groundwater geologists and petroleum engineers, is therefore appropriately named the darcy (Da).

Permeability, however, like many other complex variables, is not adequately described by a single number, even for a particular rock at a single location. This is because permeability typically shows **anisotropy:** The magnitude of permeability depends on direction. For example, the permeability of a bed of shale may vary by a factor of a thousand depending on whether the flow direction is parallel or perpendicular to the sedimentary layering, in the same way that it is much easier to split wood along the grain than across it. Recognizing such directional variability is critical in models intended to predict future groundwater availability or the migration of a contaminant. Conversely, the anisotropy of other geophysical properties of rocks—for example, the velocity at which **seismic waves** travel through them—can be used to make inferences about the nature of rocks at inaccessible depths, because the anisotropic manner in which they transmit energy is a reflection of their physical "fabric," or "grain."

Earth's magnetic field, an invisible but ubiquitous presence (and probably a prerequisite for life), is another anisotropic quantity. The unit for measuring magnetic fields is the *tesla*, a quantity named for the brilliant Serbian-American inventor Nikola Tesla (1856–1943). (Tesla, a rival to Thomas Edison, was an electromagnetic wizard who designed the mammoth hydroelectric power station at Niagara Falls after a commission headed by the omnipresent Lord Kelvin awarded the project to Tesla's patron and employer, the Westinghouse Company.) A tesla is a very large unit relative to Earth's magnetic field, whose principal (north-south) component weighs in at only about 1/20,000 tesla. For comparison, a refrigerator magnet is about 1/100 tesla, while medical instruments that do magnetic resonance imaging of the body have fields of 1 tesla or more. The strongest experimental magnetic field ever generated briefly reached 16 tesla, powerful enough to steer subatomic particles at speeds at which Einstein's relativistic laws of physics start to apply.[1] Exposure to such a field would be very harmful to living things that have evolved in Earth's far gentler magnetosphere.

Earth's magnetic field, maintained by complex motions of the liquid iron in the planet's outer core, seems minuscule in comparison with these immensely more powerful human-generated fields, but because Earth's field is global in extent, its intensity value tells only part of the story. Though measured only in thousandths of teslas, the geomagnetic field is actually a mighty magnetic shield that protects the earth from high-energy solar particles and the **cosmic rays** that zing toward us from beyond the solar system. These types of radiation can cause genetic damage to organisms and, over time, strip away a planet's atmosphere. Fortunately, the magnetic field creates just enough anisotropy in the "permeability" of space that most of these potentially dangerous particles are rerouted around the Earth, in a manner reminiscent of the way that fast-moving water flows around a rock in a stream. The scintillations of the auroras, or northern and southern lights, are a visible manifestation of this turbulent diversion.

Disturbingly, historical records show that the overall intensity of the magnetic field has declined by 10 percent over the past 150 years, which could leave Earth more vulnerable to cosmic irradiation for a few centuries. One lesson of the magnetic field is that a low-level, global phenomenon can be far more potent than a very large but localized one, in the same way that evaporation, which occurs at the molecular scale, moves orders of magnitude more water than the world's largest water diversion schemes.

A SENSE OF SCALE

Fathoming the Earth is clearly a complicated business. Earth's myriad systems span such a vast range of temporal and spatial scales, with so many different commodities in motion, that it would be easy to get lost in the details of a particular "market." In preindustrial times, different wares were often sold by different weight measures: butter by the firkin, tub, and cask; medicines by scruples, drams, and grains; wool by the clove and the sack; flour by the stone; and coal by the hundredweight. Scientists, in contrast, are generally stingy with units and economize by defining quantities that are as versatile and universal as possible. The power of geophysics comes largely from its zoom-lens ability to focus with equal clarity on everything from crystals to planets, by identifying common physical characteristics shared by disparate systems at many scales.

For example, water, glacial ice, and rock can all be treated as **fluids**—that is, things that flow, whether they are liquid or solid. One of the fundamental descriptors of a fluid is its **viscosity**—its "stickiness" or resistance to changing its shape. Molasses is more viscous in January than in June (at least in the Northern Hemisphere). Viscosity is measured in poise (pronounced *pwahz*), named for the French physiologist Jean-Louis Marie Poiseuille (1797–1869), who developed a mathematical expression for flow of fluids at low velocities in cylindrical tubes—specifically, blood flowing through veins and arteries. Interestingly, this law is nearly identical in form to Darcy's law for groundwater flow. Water at room temperature has a viscosity of about 0.01 poise. Blood, being thicker than water, registers at about three times that value. Motor oil, a product that is sold according its viscosity, ranges from 10 to 20 poise. Lava viscosities vary hugely, depending on the temperature and composition of the magma. A Hawaiian-type basaltic lava, extruded at more than 1,800°F (1,000°C) may have a viscosity of only 100 poise, while a Mount St. Helens rhyolite, a "cold" lava at 1,500° F (800°C), may have a viscosity of 10 million poise. The difference is related mainly to the silica (SiO_2) content of the lava, and it has deadly implications: The extremely high viscosity of rhyolitic lavas inhibits the escape of magmatic gases and is responsible for Earth's most violent volcanic eruptions.

Glacial ice, whose movement is the very definition of slowness, flows perceptibly over timescales of decades, and it clocks in at around 10^{13} (10 million million) poise. Rock salt cannot manage even a glacial pace (its viscosity is about 10^{16} poise) but, by geological standards, is highly mobile. Unlike other sedimentary rocks, salt is incompressible, so it does not compact or increase in density upon burial. At a depth of a mile (1.5 kilometers) or so, it therefore becomes positively buoyant relative to surrounding rocks, and squirts upward in slow motion as a solid, forming salt domes and columns (which happen to be ideal traps for petroleum). Finally, the Earth's rocky mantle moves at the most languorous rate of all. Its viscosity has been estimated at 10^{21} to 10^{22} poise—a trillion trillion times that of water. The value is derived from rates of *postglacial rebound*, the slow upwarping of crust once depressed by the weight of thick Pleistocene ice masses, like a thumbprint slowly disappearing from a nearly baked cake. The rate of rebound reflects how fast the mantle beneath the crust can flow back into place, and in places the rate is surprisingly speedy. The northern half of Lake Michigan, for example, is tilting upward at a rate of about 1 millimeter per year, spilling slowly over Chicago. In Scandinavia, thousand-year-old slips for Viking ship stand a meter or more above sea level—a design oversight those peerless boatmen would never have tolerated—indicating a similar rebound rate.

Ground motions a billion times faster are the basis for the most famous geophysical metric, the **Richter scale** for earthquake magnitude. Unlike other scales, the Richter scale is an arbitrary one with no units and no zero value. This seems illogical until you consider the nature of an earthquake. An earthquake is caused by the sudden displacement on a **fault,** or a fracture, in subsurface rocks, typically near boundaries between **tectonic plates.** The amount of **slip** in even the largest earthquakes is only on the order of meters, but the ruptured area of a fault may extend for hundreds of kilometers. Imagine pushing the Washington Monument, on its side, with no rollers, across the National Mall. It would take a huge amount of energy to move the monolith just a small amount, because a large surface area would be in frictional contact with the pavement. Similarly, a fault slips only when the energy stored elastically in rocks on either side of a fault exceeds the frictional resistance of the fault surface. The energy released travels from the point of rupture (the **focus**, or **hypocenter**) outward in spherical waves, and when these waves reach the ground surface, an earthquake is felt. These arrive first at the **epicenter**, the point directly above the focus, but may cause significant ground motions over a very wide region. Ground acceleration may locally exceed the pull of gravity, throwing free objects into the air. Structures designed to bear static loads may fail when subjected to such jolts. Seismologists have a macabre joke borrowed from National Rifle Association bumper stickers: "Earthquakes don't kill people—buildings do."

In the 1920s, Cal Tech seismologist Charles Richter recognized that understanding the physics of earthquakes required a consistent method of quantifying their effects. His greatest contribution to the field of earthquake research was to have thousands of copies of a simple but robust seismometer distributed around the globe, establishing the first worldwide seismic network.

The most basic seismometer is just an inertial mass on a vertical spring, in a box that is anchored or buried in the ground. When the ground moves up or down, the mass will stay relatively fixed. The relative motion between the ground and the mass can be recorded—as a **seismogram**—on a rotating roll of paper if a pen is attached to the mass.

Today's seismometers are electronic, but the fundamental design is the same. Richter based his magnitude scale on the response of his original instrument to ground motions of different intensities. Somewhat arbitrarily, he defined the **magnitude** of an earthquake as the logarithm of the tallest "squiggle" on a seismo-gram, measured in thousandths of a millimeter, at a distance of 100 kilometers from the epicenter of the earthquake. A magni-tude 1 earthquake was the smallest detectable with Richter's orig-inal seismometer; today's more sensitive instruments can measure seismic events with negative Richter magnitudes. The largest earthquake ever recorded, a magnitude 9.5 event that oc-curred in May 1960 off the coast of Chile, caused seismometers around the world to go off-scale.

The amount of energy released in such an earthquake is im-mense. Each integer step on the Richter scale represents a tenfold increase in the amplitude of the record on a seismogram, but a *thirty*fold increase in energy. A magnitude 9 earthquake, there-fore, has 900 times more energy than a magnitude 7 (which is still considered a sizable and dangerous earthquake). As devastating as they are from a human perspective, earthquakes are geologi-cally constructive, an insight Charles Darwin had when he expe-rienced an earlier Chilean earthquake while on the voyage of the *Beagle* in 1835. After the earthquake, he observed that the land had been lifted several feet above its earlier position. He recog-nized that thousands of such events could be the "elevatory force" responsible for the peculiar occurrence of "upraised seashells along more than 2000 miles on the western coast."[2]

Although earthquakes build mountains, seismic energy release is invariably translated into units of destructive force—equivalent kilograms of explosives. By this measure, the great Chilean earth-quake of 1960 released energy equivalent to 56 trillion kilograms of explosives—about one million times more powerful than the atomic bomb dropped on Hiroshima. Some of this energy was transferred into a giant tsunami (a seismic sea wave) that struck Japan twenty-two hours later. The earthquake also excited one of Earth's intrinsic vibrational modes, causing the planet to ring, or hum, for days at the basso profundo frequency of one cycle every fifty-four minutes—more than a million octaves below middle C.

THE IMPORTANCE OF BEING ERRONEOUS

Paradoxically, even the largest earthquakes can be traced to relatively small stresses and their effects on tiny flaws in rocks. **Stress,** in the technical sense, means anisotropic (directionally dependent) pressure. Rocks are almost infinitely strong if they experience uniform pressures in all directions. Such a state is called *hydrostatic,* since it is similar to what one experiences underwater. At the base of the Earth's continental crust, the hydrostatic (or, more accurately, *litho*static) pressure due to the weight of the overlying rock is approximately 10 kilobars—about 10,000 times greater than atmospheric pressure at the surface. While this seems an immense value, rocks at that depth will not deform (change their shape) as long as the horizontal pressures are of equal magnitude. If, however, there is even a 1 percent (0.1 kilobar) *difference* in the magnitude of pressure felt by the rock in different directions (typically as a result of tectonic movements), the rock will fail by either brittle fracture or **ductile flow**, depending on its temperature and composition. This weakness under anisotropic stress arises because rocks and minerals are full of minute defects (microcracks) that greatly diminish their aggregate strength.

In the shallow subsurface, rocks are comparatively cold and brittle, and their strength is limited by the existence of countless microcracks. Like the ends of ladders or runs in stockings, the tips of these minuscule lentil-shaped cracks are points of high stress. The load that would normally be distributed across the area of the crack (or through the knitted fabric of hosiery) is concentrated at the ends of the flaw. When rocks are subjected to significant anisotropic loads, microcracks that lie in orientations where the stress concentrations are highest will begin to propagate and coalesce into larger fractures, sometimes leading to catastrophic faulting at a massive scale. (The same phenomenon caused a horrifying air disaster in 1988, when microcracks in the body of an aging airplane suddenly coalesced and the fuselage tore off at a height of 24,000 feet as it was descending to land in Maui, Hawaii.)[3]

Microcracks are also the reason that rocks and the materials derived from them—brick, concrete, glass—are always far stronger under compressive loads than under tensile (stretching) ones. When a rock is subjected to compression, a statistical majority of the microcracks in the material will be closed, increasing the effective area of frictional interaction across the crack surface. Under tensile stresses, however, most of the microcracks will tend to be opened, reducing their area of frictional contact and leading to failure at much lower differential stresses than occur in compression. This is the principle that makes arched windows and doorways necessities in tall stone buildings. Unlike a rectangular opening, which is stretched downward in the middle, an arc is loaded compressively at every point by the overlying weight of the edifice—a fundamental physical property that the ancient Romans put to good use.

Deeper in Earth's crust, temperatures are high enough that microcracks are typically healed, or annealed, by recrystallization, but rocks at these depths are made even weaker by other, tinier defects called **dislocations**. At these sites in a crystal lattice, the stacking of the constituent atoms went awry. Imagine a group of thirty people asked to arrange themselves spontaneously into a compact and nearly square rank-and-file array. If half the people in the room set about this task thinking of five rows (and six columns) and the other half think of six rows (and five columns), the resulting array will be somewhat irregular, with a half-row and a half-column that end somewhere in the middle of the group. To remain as compact as possible, the continuous rows and columns must bend around the ends of the partial ones. This is very similar to one common type of dislocation, in which an extra plane of atoms ends in the middle of a crystal. Natural crystals are full of these minute mistakes. Like the tip of a microcrack, such a dislocation is a site of concentrated stress, because adjacent rows in the crystal lattice must bow around it, and atoms are displaced slightly from their ideal positions. If it were not for dislocations, minerals would be an order of magnitude stronger than they are.

Dislocations, like microcracks, migrate under anisotropic loads, limit the macroscopic strength of rock, and ultimately dictate the height and fate of mountain belts (just as tiny genomic copying errors are the arbiters of biological evolution). In a phrase, dislocation, dislocation, dislocation. Never underestimate the power of the Lilliputian.

Making retroactive measurements

Taking Earth's pulse and monitoring its other vital signs in real time has taught us much about the way the planet works. Instrumental records, however, are necessarily short and do not give us any sense of how the variations we detect over years or decades compare with the magnitude of changes in the geologic past. Should we be alarmed, for example, that the strength of Earth's magnetic field has fallen by 10 percent in less than two centuries? And are human contributions to atmospheric greenhouse gases really significant enough to affect global climate? Just as a doctor can better understand his or her patient's current condition by reading the person's health records, we need to consult Earth's files for perspective on its present status. But where are those files kept? And how do we read the sometimes illegible shorthand in which they are written?

Among the invisible secrets many rocks keep is a record of the magnetic field that existed at the time they formed. As an igneous rock like basalt cools from the molten state, tiny grains of iron-rich minerals (e.g., the iron oxide magnetite and the iron titanium oxide ilmenite) become aligned in the orientation of the magnetic field at the rock's particular geographic location, just as iron filings align themselves in bushy bundles in the presence of a bar magnet. These crystals not only point toward the direction of the north magnetic pole at the time but are also inclined from the vertical in accordance with the latitude at which they formed. That is, magnetic field lines are horizontal near the equator and grow steeper and steeper as one approaches the magnetic poles, where they are vertical. (During my own fieldwork in northern Ellesmere Island, in arctic Canada, which is very near—in fact a little north of!—the north magnetic pole, we had to counterweight the needles on our compasses with copper wire because they would otherwise have pointed straight down.) When the magnetic field is weak, the iron-rich minerals in a cooling magma develop a weaker statistical alignment than they do when it is strong, in the same way that a marching band with a lax leader typically has wavy rows while one under the eye of a strict director forms a perfectly ordered array. The total intensity of the remnant magnetic field preserved in a rock by the alignment of iron-bearing minerals can be used as a proxy for the strength of the Earth's magnetic field at the time of the rock's formation.

The most temporally continuous record of the history of the Earth's magnetic field is preserved in the basaltic rocks of the world's ocean floor. This magnetic archive was discovered quite accidentally as a by-product of defense-related mapping of the seafloor during the Cold War. To find new methods for concealing its submarines—and detecting Soviet ones—the U.S. Navy towed magnetometers behind ships during the 1950s. Geophysicists who were involved with these cruises became intrigued by the unexpected patterns that began to emerge as the ship-track magnetic data were compiled on military charts. When the Earth's present-day magnetic field was subtracted from the magnetometer readings, the maps revealed wide stripes, like giant UPC codes, delineating seafloor rocks with positive and negative magnetic field intensity values, or anomalies. (In the narrow scientific sense, an anomaly is the deviation of a measurement from a datum, in this case the modern magnetic field.) The magnitude of the anomalies was small—less than a microtesla (about 10 percent of Earth's total magnetic field)—but the sharpness of the boundaries between positively and negatively magnetized rocks was striking, as was the highly symmetrical arrangement of the stripes on opposite sides of the newly mapped ocean ridges.

These stripes were known for five years before their significance was understood. Then, in 1963, Cambridge geophysicists Frederick Vine and Drummond Matthews proposed that these bands represented a horizontal record of reversals in the polarity of Earth's magnetic field, imprinted on basaltic rocks pushed outward from the midocean ridges by **seafloor spreading.**[4] This idea was not only one of the most important contributions to the emerging theory of plate tectonics but also a radical new view of the Earth's magnetic field. Evidence that the planet's north and south poles had switched places repeatedly in the past suggested that the dynamics of the Earth's outer **core**—the source of the magnetic field—were far more complex than previously imagined.

Hundreds of magnetic **polarity** reversals have now been documented and dated from volcanic rocks both on the seafloor and on land. In fact, geomagnetic reversals have become important reference points on the geologic time scale. Yet the exact mechanism by which a magnetic reversal occurs remains incompletely understood. The magnetic field is an electromagnetic dynamo that sustains itself through positive feedback. In a chicken-or-egg scenario, the flow of liquid iron in the outer core together with the magnetic field generates an electrical current, which maintains the magnetic field, which regenerates the current, and so on. If it weren't so, the Earth's magnetic field would have decayed away in tens of thousands of years; that it has persisted for billions of years is remarkable. Albert Einstein in fact once identified Earth's long-term maintenance of its magnetic field as one of the great unsolved problems of physics.

While the dynamo theory is conceptually fairly simple, to model it quantitatively is, as mathematicians say, nontrivial. ("So many dynamos!"is a favorite palindrome of core modelers). But in 1995, after thousands of hours of computational time on a Cray C90 supercomputer, a magnetic reversal was simulated in a virtual version of Earth's outer core dynamo.[5] After running in one polarity mode for 40,000 years of model time, the virtual core developed anomalous patches of oppositely polarized matter in both the Northern and Southern Hemispheres, resembling the black and white dots on the oppositely colored fields of the yin-and-yang symbol. When these patches reached a critical size, the entire field flip-flopped, and north became south. The event was almost instantaneous in model time, consistent with geologic evidence that reversals occur over periods of only a thousand years or so.

In southeastern Oregon, one remarkable basaltic lava flow (long ago hardened to rock) apparently recorded the geologic moment of a magnetic reversal 16 million years ago.[6] A lava flow a few meters thick will cool from the top down in a matter of days or weeks, and under a stable magnetic field, the surface and the interior of the flow would preserve the same magnetic signature. Different levels in this particular flow from Oregon, however, record different magnetic intensities and directions. Unless something perturbed the alignment of magnetic minerals in the rock sometime after the lava cooled, the best explanation for the vertical variability in the magnetic signature is that as the lava solidified, the magnetic field was fluctuating dramatically in direction and falling rapidly in magnitude—by as much as 3 degrees in direction and 0.3 microtesla per *day*. A well-defined new north pole direction finally emerges a few lava flows higher in the stack, which may represent decades, centuries, or longer, but such short timescales are impossible to resolve with isotopic dating techniques.

The consensus is that geomagnetic field reversals probably happen on timescales of a millennium or less, short by geologic standards, but long by biological ones. During the transition period, the magnitude of the dipole (north–south) component of the field almost disappears, leaving only a much lower intensity, noisy field until the new poles establish themselves. The rapid decline in geomagnetic field intensity since 1850 could be a precursor to the next magnetic back flip. Since the last reversal occurred some 780,000 years ago, we have no direct experience of the potential effects of the absence of a strong dipole field on the atmosphere and biosphere. Happily, none of the many well-dated magnetic reversals of the past are associated with major extinction events, which suggests that greater cosmic ray bombardment from the weakening of Earth's protective magnetic field has not been extreme enough to be chronicled in the fossil record. And apparently, animals that use the magnetic field for navigation find ways to avoid getting hopelessly lost when there is no north and south.

The effects on technology, however, could be more severe. Satellite communications and electrical grids could be more susceptible to solar flares. On the other hand, the northern and southern lights could be more spectacular than ever. The literally abysmal

2002 sci-fi film *The Core* depicts global electromagnetic cataclysm as the geomagnetic field falters. Fortunately, the planet is saved by a hardy team of "terranauts" who burrow into the center of the Earth and set things right. The results of a real magnetic reversal will probably be less disastrous—but also less preventable.

TINY BUBBLES

Unlike the core, Earth's atmosphere and climate system are well within reach of humans. To assess the potential effects of human-induced climate change, we need to find natural, high-resolution meteorological records that can provide information about climate on time scales commensurate with economic planning and policy making. Although there are many valuable archives of local paleoclimate information in tree rings, stalactites, and even pack-rat middens, the very best records have been kept by long-lived polar ice caps.

For more than 200,000 years, the Greenland and Antarctic ice caps have acted as repositories for whatever has been in the air at the time.[7] The gases and particulate matter captured in the snow each year at the poles is the natural equivalent of a magazine's year-end retrospective issue. Far away volcanic eruptions, dust storms, and nuclear tests are all chronicled in these global data banks. Bubbles in the ice are like miniature vials containing samples of atmospheric gases from particular times in the past. Concentrations of naturally occurring gases like carbon dioxide and methane as well as anthropogenic newcomers like chlorofluorocarbons can be tracked on an almost yearly time scale.

These gas archives preserve remarkable detail. As a percentage of the total composition of the atmosphere, these gases seem trivial; both carbon dioxide and methane are measured in hundreds of parts per million by volume (ppmv). But even in small concentrations, these gases are highly efficient at keeping solar energy trapped near Earth's surface. While the tiny bubbles in ancient ice reveal considerable natural variability in the levels of these gases in the past, they also show, startlingly, that the *rates* of human-induced changes are almost unprecedented from a geologic perspective. For example, from the peak of the last Ice Age (about 18,000 years ago) to the beginning of the industrial revolution (around 1800), carbon dioxide concentrations increased at a rate of about 0.004 parts per million (ppm) per year. From 1800 to the present, the rate of increase has been greater than 0.6 ppm per year—more than one hundred times faster.

Such observations clearly document the relative magnitude of anthropogenic changes—those made by humans—on atmospheric chemistry, but they don't directly answer the most basic weather question: What were temperatures like in the past? The current temperature of the ice itself does not record conditions at the time it fell as snow, although temperature variations with depth do provide some general information about the thermal history of the ice mass, just as you might be able to tell how long a frozen roast had been in the oven by sticking a meat thermometer into its interior.

To get more detailed information, we have to look for temperature *proxies*—indirect indicators of past temperatures. The key is to identify a temperature-sensitive process that would leave a measurable record in the annual layers of ice. The process would be something akin to the widths of tree rings but nonbiological and less dependent on other variables (e.g., tree growth can depend not only on temperature but also on precipitation and light availability).

One such process is the sorting of oxygen and hydrogen isotopes. Such sorting occurs as water vapor moves from the tropics to the poles by way of repeated precipitation and evaporation cycles. All water is H_2O, but some water molecules are heavier than others because both oxygen and hydrogen have stable isotopes—variants with extra neutrons—that are a little weightier than the normal varieties. All oxygen atoms have eight protons and most also have eight neutrons, which gives them an **atomic mass** of 16 (in symbolic notation, ^{16}O). But some have nine or ten neutrons and masses of 17 or 18 (^{17}O and ^{18}O). Hydrogen similarly comes in the ordinary form, as single proton (^{1}H, or simply H) and as the heavyweight deuterium (^{2}H, or D) with a proton and a neutron (deuterium's name refers to its atomic mass of two). So-called heavy water, with unusually high concentrations of deuterium, is deliberately manufactured for use as a moderator in nuclear power plants—the extra weight of the water molecules slows down the neutrons zipping around the reactor and helps to sustain the fission process.

Nature has its own means of producing water with a range of molecular weights. As Earth's atmospheric circulatory system moves water vapor from low latitudes to high, the water condenses and evaporates again many times. Each time, the heavier isotopes will tend to go into the liquid phase while the lighter ones will be "selected" for the vapor phase. By the time water vapor reaches the poles, it is always isotopically lighter than it was in the tropics, with more ^{16}O than ^{18}O and more ordinary hydrogen than deuterium. Stable isotope ratios in glacial ice can be used as paleothermometers because the sorting or partitioning of the isotopes during evaporation and condensation becomes more pronounced as temperatures drop. This is perhaps analogous to the way the Cold War kept the East Germans and the West Germans strictly segregated, even though they were culturally the same people. By comparing ^{18}O-to-^{16}O or D-to-H ratios of recent polar ice with those in lower, older layers, geologists can make quantitative inferences about past temperatures, not only at the poles but globally, since the isotopic values in the ice reflect large-scale atmospheric processes. (From the standpoint of industrial efficiency, this means that heavy-water plants should use rainwater from low latitudes, during years when global temperatures are low.)

What the ice core paleothermometers tell us unambiguously is that global temperatures are strongly correlated with greenhouse gas concentrations—even though these vary by seemingly trivial magnitudes of parts per million—and that both carbon dioxide and temperature have had complex oscillations. We can see 100,000-year undulations related to the cyclical variation in Earth's orbital radius, the 26,000-year cycle caused by the planet's equatorial bulge, as well as many higher-frequency fluctuations related to ocean circulation and other phenomena. Glaciologist Richard Alley has compared Earth's climate variability in the last few hundred thousand years to "playing with a yoyo while bungee jumping off a roller coaster."[8] Importantly, however, the oscillations seem to have upper and lower limits—the bungee cords and yoyo strings have never snapped. Every nadir has a corresponding zenith, every plunge a rebound. We now have access to our patient's past records and need to make sense of them. Their complexity humbles even the best diagnosticians.

Stable isotopes in ice cores provide by far the highest-resolution records of past climate, but even the oldest layers of ice date from only the last third of the Pleistocene epoch or Ice Age (during which there were in total at least twenty cycles of ice sheet advances and retreats). Geologists have been very resourceful about finding other climate proxies that can take us further back into the past. Deep sea sediments, for example, contain several distinct types of climate signals. Like polar ice, sediments on the seafloor are repositories of global detritus. In waters far from continental margins, the rates of sedimentation are very low and the sediment that does accumulate can be considered a well-mixed sample of what is in the global ocean at a particular time. The shells of microscopic marine organisms provide several types of paleoclimate information. Various plankton species tolerate only limited temperature ranges, and their fossils provide a direct record of water temperatures during their lifetimes. In addition, trace element and oxygen isotope ratios in the shells of microorganisms provide a paleotemperature record comparable to that in ice cores, but with millennial, rather than annual, resolution. On land, pollen in lake sediments, the chemistry of ancient spring deposits, and the shapes of fossilized plant leaves have all been scrutinized for what they might reveal about the weather one day in deep time. What they all recall is variability on every time scale, high-frequency wiggles on waves riding swells, the characteristic sound of the terrestrial philharmonic.

STRETCHY COASTLINES AND IMPERIAL MICROBES

Ironically, as our capacity to fathom the Earth in its present and past states has grown more sophisticated, the very concept of measurement has become unexpectedly problematic. The most profound insight we may have gained from trying to measure the Earth may be that our measurements are never absolute, and only contextually accurate. In a 1967 paper provocatively titled "How Long Is the Coastline of Britain?" Benoit Mandelbrot of IBM gave a name, *fractal geometry*, to a branch of mathematics that could embrace the dimensional depth of the natural world.[9] Mandelbrot's point was simple: If you use a very long stick to measure a coastline, you will capture the broadest arcs but miss the fjords, firths, and coves, and you will conclude that the coastline is not terribly long. As you use shorter and shorter rulers, however, the coast actually stretches. Mandelbrot named such stretchy features

fractals because they do not fall neatly into Euclid's categories of one-, two-, or three-dimensional features but are better described as having *fractional* (noninteger) dimensions. An involuted coastline is something more than a one-dimensional line but something less than a two-dimensional plane.

Mandelbrot estimated the fractal dimension of Britain's west coast to be 1.25, while Africa's simpler, more Euclidean coast has a value closer to 1.0. The difference reflects the distinct tectonic and climatic processes that shaped the two shores. Similarly, topographic land surfaces are more than planes but less than three-dimensional solids. So depending on their ruggedness— which is the collaborative work of rock and water over time— landscapes typically have fractal dimensions between 2.0 (North Dakota) and 2.4 (New Zealand).

A fractal dimension is a metameasurement, a **scaling law** describing how much entities of different magnitudes contribute to a larger system. Decades before Mandelbrot, Charles Richter and his colleague Beno Gutenberg showed that earthquakes around the world follow a remarkably consistent global scaling law, a well-defined inverse relationship between earthquake magnitude and frequency: With each integer decrease on the Richter scale, there is a tenfold increase in the number of earthquakes that occur annually. On average, there is one magnitude 8 event, ten magnitude 7 events, a hundred magnitude 6 events, and so on, each year. If we consider this from an energy standpoint, the smaller earthquakes account for a significant fraction of the total seismic energy released each year. The one million magnitude 2 events (which are too small to be felt except instrumentally) collectively release as much energy as does one magnitude 6 earthquake. Although the larger events are certainly more devastating from a human perspective, they are geologically no more important than the myriad less newsworthy small ones.

Fault-controlled mountain belts like the northern Alps, the Canadian Rockies in Alberta, and the Appalachian valley and ridge also obey scaling relationships of this kind. Mechanically, these fold-and-thrust belts, produced by tectonic collisions, are like a wedge of snow in front of a plow. Depending on the friction at the base of the wedge and the strength, or internal friction, of the snow itself, the wedge will have a particular angle of taper. The combination of high basal friction and low internal strength (think of powdery new snow) will result in the stubbiest wedge. Low basal friction and high internal strength (crusty old snow) will form the thinnest taper. If the physical conditions that govern the frictional variables remain constant as the tectonic snowplow moves forward, the taper angle is maintained at a consistent value by the fracturing, faulting, and folding at every scale. Such a system is in a state of self-organized criticality, or dynamic equilibrium. The system maintains an overall external configuration dictated by physical laws, but energy and mass flow through its interior in much the same way that a vortex—for example, a tornado or a whirlpool—maintains its form even though the air or water molecules that constitute the vortex are not the same from one moment to the next.

Like a pile of sand stacked at its **angle of repose**, a growing mountain belt is everywhere at the point of failure, and it maintains its shape by making continuous adjustments at every scale. The largest structures are certainly the most spectacular, defining the sublime peaks and serene valleys, but if we sum the contributions of structures at all scales, we see that smaller folds and faults account for a significant proportion of the crustal thickening involved in the construction of mountain ranges. Structural geologists (like me), who study the architecture of mountains, knew this years before the advent of fractal mathematics and have long chanted the mantra "Small-scale structures mimic large."

Such "self-similarity" is an essential property of fractals, which tend to look the same, that is, similar to themselves, at many different scales (this is why geologists take so many pictures of their lens caps and Swiss Army knives). The fractal nature of biological systems has also been recognized for years. Even the eighteenth-century satirist Jonathan Swift was aware of it:

> *So, naturalists observe, a flea*
> *Has smaller fleas that on him prey;*
> *And these have smaller still to bite 'em;*
> *And so proceed ad infinitum.*[10]

We can use this property to our advantage; it implies that small systems are microcosms of the bigger ones in which they are nested, and that we can learn something about large, unwieldy systems by studying their miniature counterparts. But this also means that we tend to select an arbitrary—and usually human-sized—scale as the standard for understanding features that are actually much deeper. Even when we know about the larger and smaller "fleas," it is difficult to hold them all simultaneously in mind.

Sometimes, it's entirely reasonable to choose a scale for inquiry and squint at a system from a distance to see the aggregate result of processes occurring at smaller scales. Macroscopically measurable properties are often composite results of the unseen actions of minute legions. The bulk viscosity of fast-flowing fluids, for example, is the summed effect of miniature maelstroms, vortices, and eddies. Lewis F. Richardson, a British fluid mechanicist and pioneer in weather forecasting, took inspiration from Swift's famous epigram in his own ode to fluids:

> *Big whorls have little whorls*
> *That feed on their velocity,*

And little whorls have lesser whorls
And so on to viscosity.[11]

But in other cases, by forgetting about the scale-transcendent nature of a fractal phenomenon, we lose all sense of proportion—and miss the very characteristic that defines the system. Ecosystems are fractal, self-organized entities whose scaling relationships reflect the flow of energy from the environment to photosynthesizing **primary producers** to herbivores to carnivores. The numbers of individuals in each of these **trophic levels** (e.g., herbivore, insectivore, carnivore) is strictly limited by the contributions from lower tiers, and so as a rule, as the size of organisms increases, their numbers decrease. Interestingly, the coefficients that describe the relationship between body size and population seem to be uniform across marine and terrestrial ecosystems.[12] This relationship suggests that biological productivity in all ecosystems is bounded by the same "force field." That is, productivity is shaped by the same inviolable laws of energy transfer in much the same way that the geometry of mountain belts is governed by fundamental mechanical principles.

Big meat-eating animals are necessarily rare because their per capita energy requirements are so great. On the other hand, if lions, tigers, and bears are suddenly removed from the top of an ecosystem, their erstwhile prey (deer, rabbits) will run rampant, consuming all available vegetation and toppling the entire ecosystem in a trophic cascade. Systemwide instabilities can arise from any level, however, bubbling up from below as often as they tumble down from above. The large and the small are thus in a dynamic state of mutual constraint and interdependence. For this reason there is no single appropriate scale at which to investigate the biosphere. Organisms at all scales contribute to the maintenance of the whole, in ways appropriate to their positions in the hierarchy.

If there is any single seat of power in the global biological hierarchy, it is at the level of the single-celled primary producers, the industrious autotrophs (self-feeders). These primarily bacterial organisms are highly successful entrepreneurs that learned early on how to draw energy directly from sunlight (via **photosynthesis**), and in some cases from chemical energy from rock surfaces and mineral springs. They are the Old Ones that have persisted in an unbroken lineage for at least 3.5 billion years and constitute the link between the nonliving and the living Earth. In contrast, the reign of any particular king of the beasts—whether tiger, tyrannosaurus, or trilobite—is always short and comparatively ineffectual. Organisms high in the food chain play only bit parts in the great biogeochemical cycles that maintain planetary homeostasis. Microorganisms are the true mediators of atmospheric and oceanic chemistry and have been for eons.

Humans naturally tend to focus on macroscopic biota and are able to feel greater compassion for whales and pandas than for less charismatic microflora and fauna. Though largely unnoticed, single-celled life is everywhere on Earth—not only in cozy and hospitable places like the sunlit seas but also at sulfur-spewing volcanic vents on the lightless ocean floor, in cold dry Antarctic rocks, in slabs of sea ice, in nearly boiling hot springs, and in brine pools. A quarter teaspoon of deep-sea sediment contains a billion bacterial cells, and according to some estimates, these mucky microbes alone may constitute 10 percent of the living carbon in the biosphere.[13] Our connection with the bacterial biosphere is even more intimate and profound. According to pioneering microbiologist Lynn Margulis, "fully 10 percent of our own dry body weight consists of bacteria, some of which, although they are not a congenital part of our bodies, we can't live without."[14] In fact, a healthy human body has more bacterial cells than animal cells (bacterial cells are far smaller). Our own bodies are in some ways microcosms of the biosphere as a whole.

We need to bear in mind, however, that the present biosphere is only the latest variation on a theme the Earth has been humming for more than 3 billion years. The fossil record tells us that many global biospheres have come and gone over geologic time, sometimes dissolving slowly from one tableau to another, and other times jarred by cataclysm into a new configuration. To view nature as having achieved some ultimate and perfect balance is to underestimate grossly its subtlety and richness. Natural systems are remarkably robust precisely because no regime is permanent and no equilibrium is absolute; all components are subject to incessant inspection, culling, and replacement. The most casual look at the fossil record reveals that countless species and entire lineages have gone extinct. Yet these former biospheres are alike in some essential ways. For example, an ambitious statistical undertaking called the Paleontology Database Project, a census of all extinct organisms, suggests that global species diversity has been approximately constant since at least Ordovician time (about 475 million years ago).[15] This remarkable result suggests that the structure of the natural economy—the number of environmental niches or "jobs" available to organisms, and the relationships between producers, consumers, scavengers, and recyclers—has remained about constant over that time, even though the particular species filling those roles have changed many times over.

LAWMAKERS OR OUTLAWS?

Humans, through ever-expanding agricultural technologies, seem to have become the first species exempted from the trophic scaling laws that have shaped the biosphere for at least half a billion years. As top-o'-the-food-chain carnivores, we are unnaturally numerous. To put this in perspective: If seismic scaling laws, which are related to the energy release by the **litho-**

sphere, suddenly changed in the same way that humans have altered the energy flow through the biosphere, there would many earthquakes the size of the massive 1960 Chilean event *every year*. We could become fractally correct again if all 6 billion of us became vegetarians, but such widespread contrition for breaking scaling laws seems unlikely. To satisfy our ravenous collective appetite for meat, we have experimented with brave new trophic architectures, including turning cows into cannibals. Bovine spongiform encephalopathy, or mad cow disease, which is linked to the practice of feeding cattle the brains of their brothers and sisters, may be a warning about the hazards of trying to rewire the circuitry of life.

But is it intrinsically dangerous to change the shape and organization of the biosphere and the rules that govern it? Such manipulation is, after all, the essence of agriculture, the practice that gave rise to civilization itself—that led us from hunting and gathering in the wilderness to farmsteads and food courts. We owe our success as a species to finding clever ways to channel the flow of calories into ourselves, whether by stealing honey from beehives or applying fertilizers to fields. Our folly is to think that the rest of the system (bees, bears, bacteria) will take no notice and carry on as before. The geologic record suggests that whenever the ecological rules have changed significantly—for example, when oxygen produced by photosynthesizing organisms began to accumulate in the atmosphere, or when the first predators devoured their unsuspecting prey—old hierarchies are overthrown and a period of anarchic readjustment has invariably followed.

The scope and timescale of these reorganizations depend, of course, on the magnitude of the changes in the rules. However, our perception of the size and significance of the changes we have made is often skewed. First, the *rate* of change may be more important than its absolute amplitude in determining whether a system will be significantly affected. A ten-degree rise or fall in global mean temperature, for example, would be significant if it occurred over a millennium, but catastrophic if it happened in a decade. Each biological species has a particular reproductive interval and growth rate that place upper limits on the pace of evolution. If changes happen significantly faster than this, extinction is the inevitable result. A perturbation of a given size and rate can therefore have scale-dependent effects on the system. Rapid climate change would disproportionately harm large, slowly reproducing organisms while leaving fruitfully multiplying microbes unscathed. This observation has been demonstrated in laboratory models of ecosystems but is even more obvious in the mass extinction events of the geologic record.[16] In both the end-Permian and end-Cretaceous apocalypses, no animals larger than cats made it through the doorway into the next era.

Nonbiological components of natural systems also have internal tempos set by the physical processes and interactions that define them. The rates of ocean circulation, groundwater recharge, and chemical reactions that drive rock weathering, for example, are governed by the physical and chemical properties of water and rock. Subjected to sudden changes, oceans, **aquifers**, and rocks may not appear at first to respond, but over timescales that tend to be longer than human attention spans, they do.

More subtly, the size of an environmental rule change must be measured against the *history* of a system. Rocks and organisms have long memories, and their responses to change will always depend in part on their evolutionary heritage. Organisms "recall" their ancestors' experiences—at least those events that dictated their survival—through genetic information. If an environmental change echoes an upheaval of the past, then a species or an ecosystem may already be equipped to weather it. If, on the other hand, the nature of the change is altogether unfamiliar, the chances of surviving it may be slim.

Rocks too remember their past, which shapes present-day landscapes, water quality, and soil chemistry. Archean rocks interact with yesterday's rainwater; fractures formed in the Cambrian are reactivated in today's earthquakes. The idiosyncratic geologic heritage of a place may, in times of crisis, become unexpectedly important. For example, the meteorite impact at the end of the Cretaceous period might not have been quite so devastating had the target rocks been different. The gypsum and limestone strata at the impact site off the Yucatan coast released immense amounts of sulfur dioxide and carbon dioxide, respectively, leading first to a merciless "nuclear" winter and then a scorching global **greenhouse effect**. In this sense, the geobiosphere is not merely three- but four-dimensional, with time as the additional axis.

This inherent historicity is what distinguishes the fields of geology and biology from physics and chemistry. All electrons are the same, exempt from time and free of memory, but no rock or organism can be fully comprehended without an understanding of its evolutionary path.

Measure for measure

To fathom the Earth once seemed a simple and finite task, but measurement has turned out to be a much more elusive goal than we had expected. Earth has proven surprisingly resistant to being fathomed, at least by conventional methods. We find systems that get bigger as we look closer, and quantities that vary with direction and depend on history. We see large systems governed by the smallest phenomena in relationships that challenge our conventional notions of hierarchy and control. Perhaps the resolution to this paradox is to abandon at last the grail of ever-increasing precision in measurement, which, if pursued to the exclusion of other modes of exploration, may blind us to the inherent messiness of natural systems. The dangers of forcing nature to fit procrustean

definitions—filtering out that which seems untidy—was illustrated metaphorically in a story by Jorge Luis Borges.[17] In this
fable, the official cartographers of an imaginary kingdom are told
to draft a one-to-one map of the realm—a map so large that it
covers the entire territory and wholly obscures it. Over time, the
citizens come to view this map as reality and forget that anything
lies beneath it. Gradually, however, the map becomes worn and
ragged, finally crumbling away, and the people find themselves
wandering in an alien and incomprehensible landscape.

If creating a true-scale map of the Earth is a delusional folly,
how then should we go about fathoming our planet? Perhaps the
goal should be no more than gaining a sound sense of proportion
and what the geneticist Barbara McClintock called "a feeling for
the organism."[18] What can we say, provisionally, about the Earth
organism? It is in some ways simple and familiar; the flow of
water on and under the Earth, for example, is governed by the
same laws that direct the flow of blood through our own veins.
However, Earth's systems—atmospheric, biological, hydrologic,
tectonic, magnetic—are not only far bigger but infinitely *deeper*,
with subtle interconnections between the largest and smallest elements. The systems are also deep in the temporal sense, having
rich histories that exert strong influence on their present state.
And small phenomena can wield surprising power: A trivial deviation from sphericity causes the entire planet to wobble, raindrops and tiny flaws in minerals bring down mountains, trace
gases in the air govern climate, and microbes modulate the atmosphere. Perhaps the greatest challenge we face in attempting
to fathom the Earth is to gain a proper sense of our own size as a
human species; like spoiled children, we routinely overestimate
our importance on the planet but underestimate the destructive-
ness of our self-absorption.

4

MIXING AND SORTING

Shaken, not stirred.

— JAMES BOND

To live is to borrow.

— CHUANG TZU

.

THE WAY OF THINGS is to become tousled, jumbled, scattered, and mixed. Unruly hair rarely untangles itself. Few library books return themselves to their proper shelves. Laundry is not known to sort itself by color. Wheat does not spontaneously separate from chaff. As a result, keeping things neat and organized consumes a huge amount of human energy every day. Many of our domestic, agricultural, and industrial activities boil down to battles against **entropy**—the inexorable tendency of the universe to evolve toward maximum disorder. Counteracting this titanic tide of messiness requires immense amounts of energy.

It is somewhat surprising, then, that Earth is such an organized planet. Earth, and especially its crust, is highly refined (in the metallurgical sense), made of materials that are extremely rare in the solar system as a whole. The scores of igneous, metamorphic, and sedimentary rock types that make up Earth's crust are, in a sense, the planet's filing system—bins for specific materials that have been selectively distilled, dissolved, sieved, and sorted by geological and biological processes. Iron and nickel sank in massive volumes to the center of the Earth early in its formation, when the planet was just a blob of molten metal and rock. The silicon-rich residue formed the rocky mantle, from which aluminum, calcium, potassium, and other major elements were smelted out by volcanism to form the crust. Carbon, a relatively minor constituent of the bulk Earth and exhaled over time by volcanoes, has been preferentially sequestered in carbonate rocks like limestone. Phosphorus, present in vanishingly small amounts in the Earth as a whole, has been jealously harvested and hoarded by the biosphere. By contrast, the Moon never got very far in organizing itself, managing to generate just two distinct rock types before sliding into dormancy. With no tectonic, atmospheric, or biological mechanisms to further sort its constituents, the Moon remains a comparatively undifferentiated mass. The efficient sorting mechanisms unique to Earth are a continuation of processes that began when the solar system formed, a tale of finding needles in a cosmic haystack of stardust.

STARS OF ROCK AND HEAVY METAL

In the mid–nineteenth century, while geologists were busy mapping Earth's past—defining the boundaries and contours of the geologic timescale—Russian chemist Dmitri Mendeleev was creating the first map of the material world: the periodic table of the elements. Decades before the structure of the atom was known, Mendeleev's visionary chart arranged the elements in order of ascending mass, showed affinities between different types of matter, and even predicted the existence of elements not yet discovered. Today we understand that the periodicity that Mendeleev observed in the properties of matter has to do with the extent to which the electron energy levels, or orbitals, of atoms are filled—which elements have an electron or two to give away and which are happy recipients. Elements with similar outermost electron configurations are stacked together in columns on the periodic table and will behave similarly in the company of other elements. The snooty **noble gases**, in the far right column, have perfectly filled electron shells and refuse to interact with other lesser forms of matter.

The periodic table is a dense compendium of information about the way matter is put together and how it behaves, but the table is in some ways too egalitarian. Its neat rows and columns give equal representation to every variety of matter, from the abundant and ubiquitous to the rare and ephemeral. It's a bit like the U.S. Senate, except that each element gets just one seat. Scarce species like ytterbium and osmium (the North Dakota and Wyoming of the chemical realm), and even unnamed elements that have existed only fleetingly in nuclear reactors, have equal standing with hydrogen and helium (the equivalents of California and New York). If the seats at the periodic table were instead allotted in proportion to cosmic population, 99 percent would be reserved for these two lightest elements, perched on the top row of Mendeleev's chart. The ancient Greeks believed that there were four fundamental elements in the cosmos—earth, air, fire, and water—but in fact the composition of the universe can be summed up quite handily in just two: hydrogen and helium.

The periodic table has another shortcoming (at least from the standpoint of geologists and others who care about time and place). With the aura of a Platonic form, the table exists on an otherworldly plane (the wall of a chemistry lecture hall, perhaps) somewhere beyond the temporal and spatial realms that matter actually inhabits. The periodic table is mute about the spectacular origins, histories, places of residence, and destinies of the elements. It fails to inspire the awe its human users should feel at knowing that every atom of hydrogen, oxygen, and carbon in our bodies is simply on short-term loan from the cosmic library; was previously issued to remote stars, rogue comets, and rocky cliffs; and will soon be returned for further circulation. Finally, the periodic table doesn't acknowledge that the chart itself is an evolving entity.

The number of entries on the periodic table—the number of seats in the Senate of matter—has in fact been growing through time. The present 1 percent of all matter made of elements heavier than hydrogen and helium represents an all-time high. In the 15 billion years or so since the Big Bang, stars have steadily manufactured these elements by fusing hydrogen and helium nuclei (with one and two protons per nucleus, respectively) into heavier atoms through a process called **nucleosynthesis**. As a result, the overall "metallicity" (heavy-element content) of the universe has been slowly increasing, although it is still very small in an absolute sense. Even within the 1 percent of matter in the universe that is not hydrogen or helium, there are minority groups. The elements from lithium through iron (with three to twenty-six protons per nucleus) are, collectively, 10,000 times more abundant than all the remaining heavy elements, including the coveted metals like copper, silver, and gold, which are unusually abundant on Earth.

From this perspective, our own dense, rocky planet is a highly anomalous place, quite unrepresentative of the universe and even uncharacteristic of our own solar system. What processes could scavenge the scarce materials needed to build a solid planet from a diffuse cloud of interstellar gases? Paradoxically, assembling planets from raw star stuff begins with thorough mixing, followed only later by efficient sorting.

Scientists think that our solar system, and probably most others, began with the demise of a large precursor star (which itself may have had a family of planets) in a spectacular death scene called a **supernova**. These dramatic, explosive exits are typical only of rather large stars, those about six to eight times the size of the Sun. In its prime, such a star is capable of manufacturing not only light elements like carbon and oxygen, which are just a few fusion steps away from hydrogen and helium (and which are also forged within our own Sun), but also elements as massive as iron, nickel, and chromium. In its supernova swan song, however, a large star completely outdoes itself with alchemical pyrotechnics. In a brilliant explosion that for a few weeks releases a million times as much energy as that produced by all the stars in an average galaxy like the Milky Way, a supernova event produces a panoply of new heavy elements—including lead, gold, uranium, and other metals that on Earth have shaped human history.[1] Many of the newly minted elements are ephemeral species, unstable isotopes with short half-lives. These isotopes are simply variants of familiar elements—aluminum, for example—but their nuclear structure, with nonstandard numbers of neutrons, makes them prone to spontaneous breakdown or radioactive decay. These unstable progeny of the supernova will not themselves survive to become permanent parts of planets, but they will play a vital role in planetary construction.

A supernova explosion is so forceful that most of the mass of the erstwhile star is ejected at supersonic velocities into space. The shock wave is powerful enough to squeeze clusters of carbon atoms into tiny diamond grains. For comparison, Earth is also in the diamond-making business, but our diamonds are assembled deep in the planet's interior—it takes the pressure of 100 miles of rock to change ordinary carbon into a diamond. In other words, a supernova is a very big blast, powerful enough to trigger processes that can form a new star and, if things go right, a brood of young planets.

Our own solar system is probably a mixture of the spray from a supernova and older interstellar material that had been floating dreamily in space before being jarred into motion by the violent death of our progenitor star. We know a surprising amount about the composition of this mix—the **solar nebula** (a great name for a cocktail, I think)—from a particular group of meteorites called **chondrites**, which represent the oldest accessible, unaltered material in the solar system. Earth itself does not "remember" this stage in its history, just as people have no memory of their conception or birth. Against great odds, however, after eons of zinging through space, rare chunks of primitive solar nebula material have occasionally fallen to Earth. Against similar odds, some of these meteoritic chunks have been found by persons who recognized them as not merely extraordinary, but in fact extraterrestrial, rocks.

Chondrites are named for the tiny spherical grains, called chondrules, that give these meteorites a distinctive, pebbly structure very different from the interlocking igneous texture typical of other meteorites. The chondrules and other constituents of the chondrites are older than the planets, formed when the nebular "star guts" cocktail was still relatively homogeneous. This inference is based on the observation that the proportions of nongaseous elements in chondrites are virtually identical to the proportions of those elements in the Sun (as inferred from detailed quantitative observations of the wavelengths of light the Sun emits). The likelihood that such a match could occur by pure coincidence is comparable to the chance that two unrelated persons would have identical DNA. The most reasonable inference is that the chondrites and the Sun both sampled the same mix before it started to separate into different components. In a very literal sense, a chondritic meteorite is a piece of the Sun.

At the time that the chondrites were beginning to form, the solar nebula was still reeling from the supernova event. The gas-dust mix began to swirl, flatten, and contract into a disk with a massive central lump that would become the Sun. A fried egg cooked sunny side up is an apt mental image for the geometry of the nebula at this stage of its evolution. As the disk contracted, it rotated faster and faster (best to leave the fried egg behind now), in accordance with the principle of conservation of angular momentum, a phenomenon explicated by Isaac Newton and exploited by figure skaters performing spins. Newton was among the first to interpret the fact that all the planets "skate" (orbit) on a single plane—the plane of the ecliptic—as a dizzy memory of such a protoplanetary disk. The observation that most of the planets spin on their axes in the same direction that they orbit the Sun (counterclockwise when viewed from Earth's north pole) further supports the idea that the solar system formed from eddying whorls of primitive matter all rotating in the same direction.

Everything changed when the amount of matter at the center of the disk reached the critical mass at which nuclear fusion could begin and our Sun attained stardom. As the new star ignited, the first major sorting process began, with temperature—that is, distance from the Sun—as the organizing principle.

The manner in which the previously well-mixed nebula became separated into distinct components is similar to the *cracking*, or fractional distillation, process used to extract hydrocarbons of various molecular weights from crude oil. The oil refining process makes use of the unique boiling points for each hydrocarbon compound—benzene, kerosene, and so forth. By drawing off the vapor at each temperature interval, a chemist can extract pure substances from the previously undifferentiated mix.

In the refinery of the early inner solar system, temperatures were far hotter than they are today, and the only materials that could condense near the Sun were scarce refractory elements and compounds—those with extremely high melting temperatures—like metallic iron, calcium-titanium oxides, and magnesium silicates. This blast-furnace slag formed the planet Mercury, which broils at just 35 million miles from the Sun. A bit farther out, where it was merely scorching, iron silicate and sulfide minerals also began to crystallize, providing the raw materials for what would become Venus, Earth, and Mars. Farther still, in the more temperate regions of the disk, light and volatile elements were able to condense and stabilize, eventually forming the gas-giant planets, Jupiter, Saturn, Uranus, and Neptune. Finally, in the frigid hinterlands of the solar system, far from the radiant heat of the Sun, the solid forms of water, carbon dioxide, and methane could persist, eventually forming comets and the icy pseudoplanet Pluto.

Once the Sun was ignited, then, it sorted out the elemental bequest from its ancestors with fierce efficiency. It easily found the metallic and rocky needles in the hydrogen–helium haystack and sequestered water as ice in the outer reaches of the solar system. This separation and concentration of materials into celestial bins is a prerequisite for planets and presumably for life. Merely having the raw commodities is not enough—it would be difficult, for example, to build a house from a heap of unsorted lumber and hardware. Rather, a well-organized work site allows components to be combined again into new forms in the most versatile way. How fast did the Sun organize and assemble its materials? Again, we can look to the chondritic meteorites, the ancients that were there as planetary construction began.

Some chondrites incorporated tiny droplets of very high temperature, incandescent matter that condensed in the inner solar system soon after the Sun was kindled. These droplets, long since cooled to crystalline form, include the mineral **anorthite**, one of the most common minerals in rocks on the Earth and Moon. But geochemists who studied the famous Allende meteorite, which fell in Chihuahua, Mexico, in February 1969, noticed something peculiar about the anorthite in this chondrite.[2] Unlike normal earthly anorthite, which consists of calcium, aluminum, silicon, and oxygen, the anorthite in the Allende chondrite also contained measurable amounts of magnesium. To a mineralogist, this is like finding an uninvited and ill-dressed guest at a gala dinner. Normally, anorthite eschews magnesium because the magnesium ion is too small to fit nicely into the rank-and-file seating arrangement of the mineral's crystal lattice.

The presence of unwelcome magnesium in the anorthite in the Allende chondrite meant that either this anorthite crystallized under different rules from those applying to earthly (or lunar) anorthite, or that the magnesium somehow sneaked into the lattice under the guise of an authorized element. The second interpretation is favored by two observations. First, when the anorthite was analyzed in more detail, the magnesium was found to be residing in the places normally occupied by aluminum. Second, the magnesium was all of a particular type, or isotope, magnesium-26, which in other minerals is usually mixed with a more common variety, magnesium-24. The explanation for the stowaway magnesium is that it was produced by the radioactive decay of a short-lived isotope of aluminum (aluminum-26) that entered the crystal lattice legitimately at the time the anorthite formed and then changed its identity in situ.

The situation is a little like Cinderella's enchanted coach's being ushered through the palace gates at six o'clock and then turning into a pumpkin at midnight. Just as Cinderella had to enter the ball before the enchantment expired, the presence of magnesium "pumpkins" in the meteoritic anorthite means that aluminum-26 atoms must have entered the crystals before those atoms decayed to otherwise spurned magnesium. The presence of the magnesium-26 in the Allende chondrite places tight constraints on the amount of time elapsed between the presolar supernova (fairy godmother) event, which produced the aluminum-26, and the birth of our Sun, which is recorded by the anorthite crystals.

As unstable isotopes go, aluminum-26 has a rather short life expectancy; half of any given initial quantity will decay in about 730,000 years. A rule of thumb about radioactive isotopes is that after ten such half-lives, the amount that remains—no matter how much there was to start with—is very small. If this seems incredible, try tearing a piece of paper in half, then tearing one of those halves in half, and so on, for ten iterations. (Because of its comparatively short half-life, no primordial aluminum-26 still exists today; it was unknown until it was discovered as a by-product of Cold War nuclear experiments.) The fact that a significant amount of aluminum-26 entered the meteorite's anorthite crystals before decaying to magnesium-26 means that fewer than ten half-lives, and probably just a few million years, had passed between the supernova and the time that the anorthite crystals were being smelted out in the new solar refinery.

From a geologic perspective, the pace of events in this earliest chapter of the solar system is breathtaking. If we could peer back 4 or 5 million years into Earth's more recent past, the continents would lie in approximately their present positions and the flora and fauna would be nearly modern (although a certain lineage of bipedal apes would be living only in small enclaves in east Africa). So to go from the death throes of one star to the creation of another in the same amount of time is astounding. Only the chondrites, which opted out of planethood, remember those giddy days when the solar system was young.

The next step in planetary construction was to sweep the chemically sorted materials of the planetary disk into clumps. The sweeping process was a self-initiating and self-perpetuating dance. The chondrule-like grains of condensed nebular matter, still pirouetting at manic speed around the protoplanetary dance floor, began to stick together. As soon as any clusters formed, the gravitational pull of these objects attracted still more material, creating objects with still larger gravitational fields, and so on—a classic example of positive feedback. Little by little, over perhaps a few tens of millions of years, the nebular material coagulated, as pebble-sized objects aggregated into boulders, and eventually planetesimals that might have suited Antoine de Saint-Exupéry's Little Prince. There were probably occasional setbacks in this planetary accretion process, as slam-dancing rubble heaps with no regard for the choreography of the group collided violently with other rocky masses, launching their constituent pieces into even more chaotic orbits.

Ultimately, however, gravitational tidying-up won out over the destructive acts of rogue rocks, and the inner solar system became a safe place for young planets. We probably have Jupiter to thank for this; its massive gravity field efficiently cleared the area near the Sun of dangerous space junk. Even today, Jupiter vigilantly holds sixteen moons and thousands of **asteroids** in gravitational check. Some astronomers and **exobiologists** argue that a Jupiter-sized body (and it would almost have to be a gas giant, based on the universal scarcity of elements needed to build rocky planets) is a prerequisite for life on neighboring planets, since without such a body to act as a gravitational sweeper, an inner planet would be subject to constant, debilitating impacts.

Density is destiny

After the gravity-dominated planetary accretion process, the planets were really just rock piles in space, probably scaled-up versions of the mutually orbiting double and triple asteroids that have been viewed vicariously through space telescopes since the 1990s. (The smaller member of one such binary asteroid has been named "Petit Prince" by the International Astronomical Union. The small asteroid's "parent," Eugenia, is named for the Empress Eugenie, wife of Napoleon III, because Saint-Exupéry supposedly based the character of the Little Prince on her son Eugene.) To make a planet from a jumble of raw rock requires high-temperature baking—melting, in fact. The source of the requisite heat was once a puzzle to planetary scientists. The energy of continuing impacts and gravitational aggregation could have provided a significant amount of heat to the inner planets, but it would not have been sufficient to melt them. So again, we turn to the wise old chondrites for an understanding of those distant times.

The presence of magnesium-26 atoms (coaches turned pumpkins) within chondrites indicates that **radioactive decay** of aluminum-26 was occurring during the planetary accretion process. Radioactive shape-shifting does not happen quietly. High-energy particles and heat are given off as the nucleus reconfigures itself. So the missing heat source for the planetary kiln came from the planets themselves—the radioactive decay of short-lived isotopes like aluminum-26, the deathbed legacy of our ancestor star and the inheritance we needed to make our way on our own.

Once in the molten state, each planet could easily sift through its bequest. The heaviest components—in Earth's case, iron and nickel—simply sank to the center of the liquid planets while silicate magmas separated out and floated upward like the head on beer. This gravitational differentiation formed the cores and mantles of the planets. While we cannot directly sample the deep interiors of Mars, Venus, or even Earth (sorry, Jules Verne), we do have analog rocks that record this stage in planetary development—the iron and achondritic stony meteorites. Younger than the chondrites (which never melted or differentiated) these igneous meteorites are thought to represent fragments of the cores and mantles of ill-fated planetesimals that had differentiated and resolidified before being shattered by late collisions with other star-crossed objects. (And these are the types of meteorites used as proxies in determining the age of the Earth.) A few very special meteorites, the stony-irons or pallasites, appear to chronicle the differentiation process as it occurred. Exotic hybrids of iron-nickel alloys and lovely sea-green crystalline olivine, they record the incomplete separation of metal and silicate melts within the interior of a planet that would not reach maturity.

In the middle of its own differentiation period, Earth too seems to have experienced a near-fatal crash. The current theory for the origin of the Moon is that just when the early Earth had finished separating into core and mantle, the planet was struck in a glancing collision by a Mars-sized planet that was at a similar point in its development.[3] The result was a series of events so complex that simulations of it require hours of supercomputer run-time. These models suggest that the core of the colliding planet (dubbed Orpheus) was largely subsumed into the Earth, while its mantle, together with some of Earth's mantle, was sprayed into space just far enough that the molten matter coalesced again into a satellite rather than form a planetary ring. Yet again, we see how Earth's path to planethood involved cycles of mixing and sorting.

This rather alarming scenario for the birth of the Moon, while impossible to prove, is consistent with many puzzling characteristics of the Moon and is embraced by most planetary scientists as the best available working hypothesis. First, it explains the slightly-too-frenetic angular momentum of the Earth-Moon do-si-do, as well as the slight tilt of the Moon's orbit off the plane of the ecliptic (if it were not so, solar eclipses would be a monthly occurrence). Second, the hypothesis resolves the maddening geo-chemical paradoxes posed by the lunar samples collected during the Apollo (U.S.) and Luna (U.S.S.R.) Moon missions. Analysis of these samples showed that in some respects the Moon's rocks are very similar to their terrestrial counterparts—especially in their oxygen isotope ratios, which are as distinctive as fingerprints in identifying rocks with common origins. Yet the Moon rocks (especially the lunar basalts) are also crucially different from Earth rocks of the same types: Lunar rocks are very high in titanium and iron and are nearly bone-dry. These observations seemed to disprove all the earlier hypotheses about lunar origins; the lunar rocks were too similar to Earth's for the Moon to be a captured planet, but too different for the Moon to be a twin sister or a spalled-off clone. Instead, the Moon seems to be a mix of earthly and Orphean matter, the native and the alien. The complete lack of water in its crust reflects its twice-baked past; the vaporization of the molten outer mantles of Earth and Orpheus at the time of their collision left the Moon without a drop to drink.

The Moon's pockmarked surface reveals that large impacts continued for another billion years after its tumultuous creation, but this mother of all impacts was probably one of the last to cause a major reorganization of the celestial dance. As the short-lived isotopes produced in the supernova finally burned themselves out, the planets slowly solidified, at rates inversely proportional to their size. Mercury and the Moon froze entirely within a billion years. Mars fought off hypothermia for another billion or so. Earth's rocky mantle froze too, but even today is mobile enough in solid form to turn over, like very sticky taffy, about once every 200 million years. The only planetary region in the inner solar system that has remained entirely molten since those primordial days is Earth's outer core, the planet's magnetic dynamo.

The story of the early solar system, then, is like a ballad with a repeating chorus. Overall, it is a tale of sorting and sifting, but each verse is followed by a refrain in which the winnowing is partly undone by an event that mixes new material back into the pot. In the beginning, an ancestral solar system, perhaps an orderly place with planets and moons of its own, is turned to an element soup when its star explodes. Then a new star ignites and sorts the mess by temperature, collecting rocky and metallic chunks in inner orbits. The chunks jostle, collide, and aggregate, becoming giant rubble piles of assorted composition. These piles of rubble pool their radioactive heat and melt, which allows them to separate into dense metallic centers surrounded by rocky mantles. Just when Earth has sorted itself in this way, it is struck by a peer planet at a similar stage. The two exchange material, forming a somewhat larger Earth and a small companion moon. Both then return to the business of differentiating.

Once the planets had sorted themselves into metallic cores, rocky mantles, and primitive crusts, this segregation would seem irreversible. The layers had reached their density-based destinies. How could the buoyant upper crust be induced now to interact with the lower echelons of this stratified system? Interestingly, a growing body of evidence suggests that Earth's unrivaled stability and clemency can be attributed to the ways in which the planet maintains communication and exchange between its interior and exterior. And in almost every one of these transactions, water is involved as emissary, diplomat, shipper, or provocateur.

WHITHER THE WATER

Although Earth and its nearest neighbors, Venus and Mars, began with about the same inheritance of metal and rock, the biographies of the three planets began to diverge soon after their formation. Mars was congenitally disadvantaged by its diminutive size, first, because it lost heat more rapidly and, second, because it lacked the gravitational authority to hold onto an atmosphere. Venus, though closer in size to Earth, never developed ways to compensate for its proximity to the Sun and allowed a temperature management problem to become chronic. Only Earth developed habits of self-maintenance that have kept it looking youthful and fresh. Earth's beauty secret: Water, and lots of it. But where has this water come from? And didn't Venus and Mars get any?

Venus, Earth, and Mars were all too close to the blazing new Sun for pure water in any form to be stable at the time that materials condensed out of the solar nebula. All three planets, however, probably did receive water during this time in the mineral equivalent of a heat-proof vault: hydrous **silicate** minerals called **amphiboles**. The amphiboles are a large and illustrious family of minerals whose members include hornblende, a primary constituent of granite; nephrite, one of the components of the rock commonly called jade; and a couple of black sheep—two types of asbestos. The characteristic family trait of amphiboles is a long, lanky rodlike structure that reflects a crystal lattice in which water, in the form of the hydroxyl ion (OH^-), acts as mortar for silicon, oxygen, magnesium, calcium, and other elements that serve as the bricks of most rock-forming minerals. The chemical formula for the amphibole called tremolite, which can take the notorious needlelike form of asbestos, is $Ca_2Mg_5Si_8O_{22}(OH)_2$. This composition means for every eight atoms of silicon (Si, the essential constituent of 95 percent of all of Earth's rocks), there is a molecule of water. Although amphiboles were only a small fraction of the nebular material that condensed in the inner solar system, they did provide Earth, Mars, and probably even Venus with a considerable amount of native water, albeit locked in crystalline safe-deposit boxes. The key to releasing such water is heat and pressure; amphiboles tend to break down at depths of about 60 miles below Earth's surface. Under these conditions, the previously captive water is then released from its mineral bonds and may eventually escape through volcanic conduits to the surface as steam. So, some of the water on the early Earth (as well as Venus and Mars) is probably indigenous, boiled out of the magma oceans and belched from the volcanoes that covered the surfaces of the new planets.

But the Earth (and probably Mars) somehow acquired far more water than can be accounted for by the degassing of Earth's interior. Some water must have been imported from elsewhere, and the most likely supplier would have been comets on long sojourns far from their homes on the frigid outer margins of the solar system. Most comets have highly eccentric and elliptical orbits, and over time, many have taken one-way trips into the inner solar system, unable to resist the gravitational charm of the planets. (This was the fate of comet Shoemaker-Levy, which in 1994 was inexorably drawn toward Jupiter in a spectacular display of celestial mingling.)[4] Incredible as it may seem, as much as half of Earth's water may have been delivered in this way, although the proportion that is imported versus domestic is a controversial issue among geochemists.[5]

Whatever its origins, Earth's water is implicated in virtually every geologic process both at and below the planet's surface. The importance of water to life is obvious; in fact, water is considered the sine qua non in the search for extraterrestrial life. But the roles of water in the workings of the solid Earth are equally profound and more surprising. Often functioning as a subversive via underground channels, water alters the physical and chemical properties of rock in ways that radically affect—and in fact define—Earth's plate tectonic system.

For example, water is directly responsible for the generation of continental crust, the buoyant granitic distillate unique to Earth. Today, new continental crust forms at ocean crust recycling centers, better known as subduction zones. These zones occur where old, water-saturated seafloor, produced long before at a midocean volcanic ridge, has become cold and dense enough to sink back into the mantle. A slab of such oceanic crust descends slowly into the planet's interior at an angle ranging from about 10 to 45 degrees, depending on the slab's age and consequent density contrast with the surrounding mantle (a young, warm, and buoyant slab cannot plunge into the mantle at a steep angle). Subduction is occurring today off the west coast of South America, at the deep ocean trenches flanking the Indonesian and Philippine Islands, and in the seas adjacent to nearly all of the world's most dangerous volcanoes.

As a subducting slab experiences progressively higher temperatures and pressures, it releases water from amphiboles and other hydrous minerals such as clays (which are formed when water interacts with igneous minerals at Earth's surface). This water acts as a **flux**, which, like carbon added to iron in steel making, lowers the melting temperature of the otherwise solid overlying wedge of mantle rock. This low-temperature magma is quite different in composition from the bulk mantle rock from which it comes, because fractional melting produces a liquid that has relatively high concentrations of the elements that are the most eager to escape from the crystals in which they occur. These elements, which geochemists call **incompatible**, are typically rather large ions that would prefer to have more space than the rigid structure of a mineral lattice provides. The incompatible elements happen to include potassium, rubidium, and uranium, all of which come in radioactive varieties. As a result, the continental crust is not only different in bulk composition but also hotter—in the radioactive sense—than its progenitor, the mantle. On a planet made of stuff that is very rare in the solar system, the continental crust is made of rarer stuff still. No other planet has managed to produce such a concentrate. Earth's secret to scav-

hwestern corner of the Pacific Ocean). In other words, ocean
t dies young but gets continuously reincarnated.

we extrapolate this rate of overturn back in geologic time,
ocean floor has apparently been rejuvenated at least two
en times since the Earth formed. When Earth was younger
hotter, however, the pace of convection may have been
r, and the ocean floor may have been resurfaced more fre-
ntly. But this leads to a conundrum: If convection had been
r in the past, as most geoscientists think it was, ocean crust
ld have arrived at subduction zones at a younger average
still too hot and buoyant to be assimilated back into the
tle. This suggests that true **plate tectonics**, with rigid crustal
s, efficient recycling of ocean crust via subduction, and water-
ted production of low-temperature melts, may not have oc-
ed on the early Earth.[6] Instead, plate tectonics could begin
when the Earth had reached a degree of thermal maturity,
ably about 2.5 billion years ago (around the close of the
ean eon and the beginning of the Proterozoic). Before this,
h's mixer settings—and the extent to which surface water
stirred back into the interior—were probably different. We
ook to rocks formed in these distant times, Earth's record of
ildhood and youth, for clues.

chean rock complexes, found in the interiors of most of the
nents, including the vast Canadian Shield of North America,
est that a kind of marshmallow tectonics prevailed on the
ce of the young Earth. Two rock assemblages are characteris-
f these Archean formations. First, and generally older, are
tacularly deformed and highly metamorphosed **gneisses**
impressionistic whorls and streaks of light and dark miner-
uch rocks represent igneous intrusions that were caught up
ultiple enigmatic tectonic upheavals—collisions between
eval continental masses. This group includes the oldest rocks

enging these comparatively rare elements is to mix surface water
back into the planet's interior.

MIXED DRINKS AND METAPHORS

Let's think about mixing. I have always loved the thesaurus-like
list of verbs on blenders: stir, knead, chop, mince, puree, liquefy,
whip. These embody some essential physical facts about mixing.
First, solid components must be reduced to small particles to be
efficiently mixed with other materials. Grinding not only creates a
mixture more homogenous at fine scales, but also provides larger
surface areas across which the particles can interact with the surrounding medium (because of the high area-to-volume ratio of
small objects). Second, mixing is all about motion, getting every
part of a system to come into at least brief contact with every other
part. High-velocity, disorderly motion—what fluid-mechanics experts call turbulence—is more likely to result in thorough mixing
than is orderly motion, or laminar flow. That may be why James
Bond prefers his martinis shaken, not stirred.

The Earth doesn't have buttons with preset blending speeds.
How does it do its mixing? Mixing can occur only where there is
a flow of material from one place to another, and flows happen in
response to inequality—gradients, or spatial differences, in the
value of physical variables. Flows can be thought of as Robin
Hood processes, always robbing from the rich to give to the poor.
This is the essence of the Henry Darcy's law for groundwater flow
and Jean Poiseuille's rule for blood circulation (Chapter 3).
Groundwater, magma, and blood flow as a response to pressure
differences. Surface water flows downhill to redress differences in
elevation (potential energy). Heat flows to equalize temperatures.

The slowest kind of geological process, called **diffusion**, is driven by differences in the concentration of particular elements

from one place to another. In diffusional flow, atoms move through a stationary medium in response to concentration variations. This can be a very slow and arduous process, especially if the medium is a crystal lattice—rather like making one's way without a machete through a very densely vegetated jungle. Such solid-state diffusion in rocks proceeds more quickly at high temperatures, but even at its most efficient, diffusion cannot move atoms faster than about a twentieth of an inch per a year. Diffusion is significantly faster through a liquid (more like walking through a temperate forest) and fairly efficient through a gas (strolling across a well-clipped lawn). Heat flow by **conduction**, in which heat energy moves through a static medium from hot areas to cold (e.g., from a stove element into a cold frying pan), is also a diffusive process. Thermal conduction is extremely slow in rocks, which are notoriously poor heat conductors.

In many geological settings, however, "walking," or diffusing, through a medium may not be the only transportation option. Sometimes atoms and heat can hop a ride on a moving medium and thereby cover much greater distances. This is called **advection**, and water is often the vehicle for mass transit. A dissolved ion can travel tens of feet per year in moving groundwater and hundreds of miles per year in surface water, far outstripping a pedestrian counterpart trying to diffuse through dry rock.

Over longer timescales, rock itself can also act as a vehicle for advection. Earth's solid mantle is in a steady but incomprehensibly slow roil, turning itself over in a special type of advection called thermal convection, the motive force for plate tectonics. **Convection,** like conduction, is driven by temperature contrasts, but somewhat indirectly. Convection requires a pronounced vertical temperature gradient, with cool material overlying hotter stuff. As in a lava lamp, mantle rock is heated from below (not by a light bulb but by primordial heat from the core and radioactive

heat from the mantle itself). The key requiremen temperature-related volume change—expansi contraction upon cooling. (In a lava lamp, the c ently undergoes greater thermal expansion th This sets up a buoyant instability because the denser than the warmer underlying material. bottom rises, and the whole system stirs itself. the top of this system, moving at rates of inch happens to be about as fast as fingernails grow

Convective overturn occurs in the Earth's int of a fortuitous combination of physical variat better conductors of heat, such stirring would because the necessary temperature contrasts v If rocks didn't expand significantly when heate no density instability to drive convection. If the mantle rocks were much higher, the whole syst a halt. Finally, if the planet had a smaller inven elements, or if these had much shorter half-liv planetary lava lamp would have burned out lor mantle keeps on turning.

THE MANTLE OF POWER

The rate of convective overturn in the mantle from the age of the oldest surviving ocean crust the upper lid of the circulating system. Unlike tal crust, whose density at any temperature is mantle rock, ocean crust as it ages eventually dense enough to sink back into its source, th duction. Under Earth's current thermal condi reaches negative buoyancy when it is about 17 (the age of the oldest extant ocean crust, wl on Earth: the venerable 4.0-billion-year-old Acasta gneisses of the Great Slave Lake region, Northwest Territories, and the exquisitely exposed Isua complex of western Greenland (at 3.8 billion years, the Greenland rock is slightly younger). Then there are the granite-greenstone belts, two-tone igneous complexes composed of both continental-type rocks (the granites) and oceanic crust (the **greenstones**). Greenstones are mildly metamorphosed basalts whose morphology and color both reflect the ubiquitous presence of water even on the early Earth. These basaltic rocks commonly occur in bulbous shapes, called pillows, which indicate that the lava was extruded underwater and quenched like homemade candy into spheroidal blobs. The verdant tint of the greenstones comes from the hydrous mineral chlorite, which indicates that the rocks were altered by heated groundwater sometime after they solidified.

The geologic map of the Canadian Shield—including parts of the Northwest Territories, Manitoba, Ontario, Quebec, the Upper Peninsula of Michigan, and Minnesota's Boundary Waters region—is a boldly striped canvas on which east-northeast-oriented bands of gneisses alternate with lenticular granite-greenstone belts. The regional alignment of these bands suggests that the rock units were squeezed together by horizontal shortening, rather like the striped mounds formed if you gather the froth on a latte to one side of your mug. Although this banding suggests considerable tectonic crowding and jostling in Archean time, these ancient rock complexes do not look like the eroded roots of modern-style mountain belts and subduction zones. Archean tectonic zones lack the thin-skinned fold-and-thrust belts (like the Canadian Rockies) that indicate the presence of a rigid crust beneath a veneer of crumpled, stratified rocks. Also absent are the fingerprints of subduction: the high-pressure metamorphic rocks called blueschists (Chapter 2) and

eclogites, dense, garnet-bearing rocks formed when basalt is pushed down into the mantle.

Then at about 2.5 billion years ago, an important change apparently occurred in the thermal and mechanical character of the Earth's crust. One line of evidence for this is the Great Dike of Zimbabwe, a giant, magma-filled crack 300 miles long, several miles wide, and extending in depth through the entire thickness of the crust of central Africa. The existence of this mammoth fracture, paradoxically, indicates that the crust was strong—that it had then cooled to the point at which it could fail like any other brittle material, in a colossal crack. The basaltic rock that filled the crack and forms the dike records its age—2.5 billion years. Thus, the dike is considered the golden spike marking the Archean–Proterozoic transition. Soon after the formation of the Great Dike, the first mountain belt of modern design, the Wopmay **orogen** in Canada's Northwest Territories, and the first unambiguous subduction-related rocks, a complex of eclogites in Tanzania, appear in the rock record.[7]

So, it seems that after Earth's ebullient Archean childhood, the planet settled into its more methodical tectonic habits by Proterozoic time and has maintained them ever since. But there is a further paradox: Growing evidence suggests that near-modern volumes of continental crust (far more than survives in the exposed Archean complexes) existed 3 billion years ago or earlier, well before modern-style plate tectonics is likely to have begun.[8] First, the oldest preserved native Earth materials, the ancient zircon crystals found in Archean sandstones from western Australia, are thought to have been derived from a granitic source, which suggests that at least some continental crust existed as early as 4.4 billion years ago.[9] Second, more compelling, if indirect, evidence indicates that large amounts of continental crust formed early in Earth's history. The evidence comes from rare components of

Archean greenstone belts: high-magnesium volcanic rocks called komatiites (named for Komati, Swaziland, where they were first described). These rocks typically have long, needlelike crystals, indicating rapid crystallization of an undercooled liquid (think of the slender, filigreed ice crystals that form when water vapor comes in contact with cold window glass). These crystals give komatiites a distinctive texture called *spinifex*, after a type of tall grass that grows on the savannas of Swaziland.

Komatiite is an "extinct" rock type in that it no longer forms on the modern Earth (most of the preserved komatiites range in age from about 3.0 billion to 3.6 billion years old). With crystallization temperatures of about 1,600°C, magmas of this composition would today solidify before reaching the surface of the Earth as volcanic lavas. Komatiites therefore support the inference that temperatures in the upper mantle and crust of the young Earth were significantly higher than they are today. The trace element geochemistry of komatiites reveals other facts about the Archean mantle. Given their age, we might expect that komatiite melts would have come from a "primitive" mantle that had not been significantly drained of incompatible trace elements—the large ions that escape into melts if they possibly can and that are therefore preferentially concentrated into continental crust. But komatiites are surprisingly low in the most incompatible trace elements, which indicates that the upper mantle had already been depleted of these elements when the komatiite magmas were generated. This in turn suggests that significant volumes of continental crust, or a primitive version of it, existed by about 3 billion years ago.

The puzzling implications of this inference are (1) that the first continental crust was apparently produced by a process different from the modern subduction method and (2) that rates of creation and destruction of continental crust have been commensurate with each other over geologic time, even though continental crust is too buoyant to be recycled via subduction. Put another way, how could so much continental crust have been created in Archean time, and how could so much have been destroyed since then? Water is part of the answer to both questions.

Subduction is the principal way in which the exterior and interior of the modern Earth interact. Today, subduction involves the dehydration of old ocean crust at depths of 30 to 70 miles, depending on the temperature (and thus age) of the descending slab. Release of water that the ocean crust acquired at the Earth's surface leads to the generation of **arc magmas** (i.e., future continental crust). The basaltic slab itself is then transformed into much denser eclogite, a Christmassy combination of bright green pyroxene and deep red garnet. The high density of eclogite helps to pull more of the ocean slab into the mantle. If this did not happen, the oceanic slab would become neutrally buoyant at a shallower depth in the mantle and its descent would not be as rapid—that is, the pace of tectonics on Earth is set in part by the rate at which basaltic ocean crust can be metamorphically transformed to eclogite.

Although the formation of eclogite involves the extraction of water from subducted basalt, the metamorphic conversion to eclogite becomes very sluggish if the dehydration process is too thorough, or if it happens before the slab reaches the critical depth at which eclogite minerals can form. This is because metamorphism requires relocating atoms from old crystal structures into new ones—rather like moving people from one terminal to another at a busy airport. Shuttle buses are the quickest and surest mode of transport, but if the buses aren't running, passengers have to trudge along labyrinthine corridors and sky walks and may get to the gate too late. As long as even a small amount of water remains in a subducting slab, it can shuttle (advect) atoms in dissolved form to new locations, and eclogite metamorphism proceeds efficiently. On the other hand, if subducted basalt is heated and dehydrated before being converted to eclogite, all the atoms will have to "walk" (diffuse) to new crystal sites. This process can be so slow that eclogite will not form in any quantity, even when the rock sits at eclogite-forming depths for millions of years.[10] So, conversion of basalt to eclogite seems to require that just the right amount of water remains in the slab as the erstwhile ocean crust sinks into the mantle.

In Archean time, when the mantle and crust were hotter, less surface water would have made it into the mantle. "Marshmallow" tectonics atop a more vigorously convecting mantle would have created areas of thickened basaltic crust. Some of this tectonically buried crust would have been hot enough in the presence of water to be partly melted. This melted crust is thought to be the origin of the surprisingly large volumes of granite in Archean **shield** areas. Unlike modern subduction-generated magmas, Archean granites have chemical signatures that suggest they formed via melting of hydrous basalt at the base of such thickened crust, without the involvement of mantle material. Partial melting draws water out of rock very efficiently, and so the residual lower crust would have been severely dehydrated, too parched to recrystallize to eclogite even at depths where eclogite should have formed. This dry, "uneclogitized" crust would have been very strong and too buoyant to sink into the mantle, and it may have formed the substrate for the first true plates. Only when upper-mantle temperatures had moderated by a few hundred degrees (around the time of the Archean–Proterozoic transition) could down-going oceanic crust remain cold enough to retain its water to eclogite-forming depths. Once these conditions were met, modern tectonics—with efficient eclogite production and the generation of arc magmas from the water-recharged mantle—could finally begin. In a sense, plate tectonics required a deeper connection between Earth's hydrosphere and interior than was possible on the hot, young Earth.

WASTE MANAGEMENT

Evidently, then, large amounts of continental crust existed already in pre-plate-tectonic Archean time, formed by the melting, not of the water-recharged mantle, but of the still-hydrated lower crust. If it is not possible to subduct continental crust—the Himalaya Mountains illustrate the stubborn refusal of buoyant continental crust to be pushed down into the mantle—why hasn't the volume of continental crust been growing steadily through time? How has Earth been getting rid of the stuff? The answer is . . . raindrops.

One of the most remarkable characteristics of the Earth is that the rates of its interior (tectonic) and exterior (climatic) processes are approximately balanced. Erosion can dismantle mountain belts nearly as fast as they grow. This is a happy coincidence, and possibly a crucial one for keeping a planet on an even keel, but not all planets enjoy the moderating effects of erosion. The huge volcanoes on Mars attest to this. Like someone with a pituitary malfunction, they grew through the unchecked accumulation of lava flows, and nothing stopped the volcanoes from reaching gargantuan heights. On Earth, there are limits to growth, imposed largely by running water. No mountain is exempt from erosion, and the steepest slopes are subject to the fiercest attacks.[11]

But simply eroding a mountain belt isn't enough to remove continental crust from the surface of the Earth. Most of the sediment will accumulate in low-lying basins and on the **continental shelves**, the shallowly submerged fringes of the continents. Although the shelves are below sea level, geologically speaking they are continental (granitic and buoyant), not oceanic. Most of the classic sedimentary sequences around the world—the strata in the Grand Canyon, the limestones of Indiana, the dolomites of the Italian Alps, the oil-saturated beds of the Persian Gulf region—all represent sediments that accumulated on top of continental crust. These sediments collected either on the continental shelves or in interior basins at times when global sea level was high enough to inundate low areas in the hearts of continents (even Wisconsin once had tropical reefs). Sedimentary sequences on continental shelves are on the front lines when continents collide—for example, the folded strata of the Appalachian valley and ridge, the marine limestone at the summit of Mount Everest—but as long as sedimentary rocks are perched on top of continental crust, they are safe from subduction. The only way to send continentally derived sediment back to the mantle is to take it so far out to sea that it gets a ride on the oceanic express and goes down with a subducting slab.

Getting continental sediment onto the deep seafloor would seem difficult. As great rivers empty into the sea, they lose their gravitational energy and particle-carrying capacity and must relinquish the sedimentary souvenirs they have collected upstream. Most of this detritus gets dropped near their mouths as **deltas**. But some of the world's greatest river systems—for example, the Indus and Ganges, which begin in the high Himalayas—transport so much sediment, most of it too fine to settle near shore, that their submarine fans extend a hundred miles out onto the seafloor.

Earthquakes may also assist in shaking sediments off continental shelves and down to the deep sea. Seismically induced landslides produce a characteristic type of sedimentary deposit called a **turbidite**, named for the turbulent water-sediment slurries that form these deposits. Extensive exposures of such rocks, often strongly folded, had been mapped and described as early as the middle of the nineteenth century in mountain belts around the world, particularly the Alps, the Appalachians, and the British Caledonides. The strata, which were (and sometimes still are) referred to as graywacke or by their Swiss name, *flysch*, were remarkably alike from place to place, despite differences in age. All were uniformly bedded on the scale of inches to feet. Each bed began with rather coarse-grained material (gravel or sand) and became finer upward, indicating first rapidly flowing and then quiet water conditions. The tops of some beds appeared partly eroded or truncated, evidence of forceful scouring as the succeeding bed was deposited. Most remarkably, the sequences consisted of hundreds or thousands of such layers, with total thicknesses of half a mile or more. And finally, most flysch sequences were found in the highly deformed interiors of mountain belts. For decades, flysch posed a challenge to geologists' uniformitarian thinking, since no known process could produce such thick sequences of strata with these distinctive characteristics.

The answer was presented unexpectedly in 1929, when an unusual earthquake on the Grand Banks off the coast of Newfoundland was followed by the breakage of trans-Atlantic telegraph cables in rapid succession, from shallow water to deep. Based on the times that the cables were broken, it was calculated that whatever entity cut them was traveling as fast as about 45 miles an hour.[12] This was the beginning of the recognition of turbidity currents—powerful, density-driven slurries of sediment—as important vehicles for transporting continentally derived sediment to the seafloor. The volume of sediment transported by the Grand Banks was estimated at about 24 cubic miles (100,000 times the volume of debris removed from the World Trade Center site). After careening down the submarine canyons of the Grand Banks, the juggernaut of sediment reached the deep seafloor and spread out over an estimated 40,000 square miles (an area equal to four Vermonts).

The flysch deposits of the Alps, Appalachians, and other mountain belts apparently represent the deposits left by hundreds to thousands of such events. More recent studies have shown that the thicknesses of layers in some turbidites have scaling laws very similar to the fractal Gutenberg-Richter relationship for earthquakes (Chapter 3), consistent with seismic triggering (and suggesting that turbidites could provide information about paleoseismicity in a given region).[13]

It would be another half century after the 1929 earthquake, when Earth's plate tectonic system was discovered, before the importance of turbidites to Earth's long-term crustal recycling system would be understood. Although the Alpine and Appalachian turbidites (and all the others that geologists have gotten their hammers on) escaped subduction, most turbidity deposits probably ride the ocean floor conveyor back into the mantle. Although there is some disagreement among geologists about how much continentally derived material is subducted every year, the emerging consensus is that it is a considerable amount. Geologists speculate that the volume of material is probably very close to the amount of new continental crust produced by the magmas from explosive, subduction-related volcanoes at island arcs (e.g., Japan, Indonesia, the Philippines), and continental arcs (e.g., the Andes, the Cascades).[14]

Arc magmas provide some compelling evidence for surprisingly rapid recycling of continent-derived sediment, in the form of **cosmogenic isotopes**. These special isotopes are very much like tracers used in medical tests. They have a known origin and can shed light on regions that cannot be accessed directly. Unlike the inventory of isotopes that Earth inherited from the primordial supernova event, which have been steadily disappearing through time, cosmogenic isotopes are a renewable resource. They are constantly regenerated in Earth's uppermost atmosphere as atoms of carbon, nitrogen, and oxygen are bombarded by high-energy cosmic rays. These collisions produce a variety of short-lived radioactive isotopes, including carbon-14 (^{14}C), whose half-life of just under 6,000 years makes it a useful dating tool in archaeology.

Other cosmogenic isotopes have somewhat longer half-lives and are thus better suited for geologic investigations. One of these isotopes is beryllium-10 (^{10}Be), some of which (like carbon-14) can find its way into the lower atmosphere. From there it is "rained out" onto the land and into the oceans, where it sticks to the surfaces of continentally derived clay particles. Some of these clays eventually reach the seafloor and get subducted with the underlying ocean crust. Beryllium-10 has a half-life of about 1.6 million years, so after about 16 million years, virtually none will remain. Astoundingly, significant amounts of ^{10}Be have been detected in the emissions of arc volcanoes in the western Pacific (including Mount Pinatubo in the Philippines), indicating not only that continental materials are entering the mantle in large volumes, but that some components are being returned to the surface with remarkable efficiency.[15]

If we step back and picture the Earth as a whole, at timescales commensurate with its "circadian" rhythms (tens to hundreds of millions of years), we see a planet that has continued to practice the repeated mixing and sorting cycles that characterized its formation. The Earth continuously distills two types of crust from its interior via distinct processes at different rates, and simultaneously recycles each type apace with its production, thereby maintaining an approximately constant ratio of the two. All parts of the fabrication and recycling process are cleverly linked and powered largely by water. The destruction of ocean crust via subduction leads to the formation of continental crust through water-facilitated melting. The destruction of continental crust via water-driven erosion ultimately replenishes the mantle for the next round of ocean crust production. Efficient, sustainable, robust, and elegant, the system would win top honors in an industrial design competition.

MAL DE MER

The same aesthetic sensibility can be seen in Earth's atmosphere and ocean, where circular motifs are ubiquitous. Convective mixing is especially important in the air-water system, as a means of minimizing the inevitable disparities between high and low latitudes and shallow and deep waters. But unlike the slow, steady convection within the solid earth, which has proceeded uninterrupted since Earth's formation, the surface cycles of air and water are more mercurial. Prone to stormy tantrums, the Earth's air and water create the planet's notoriously unpredictable weather (without which human conversations would be reduced by at least half). But even if tomorrow's weather cannot be predicted with certainty, next June's weather can be guessed with reasonable confidence. That is, despite their volatility (literal and metaphorical), Earth's atmosphere and hydrosphere are not generally fickle or capricious over longer timescales. Year after year, ocean currents retrace consistent courses; atmospheric circulation occurs on a clockwork schedule. Daily and seasonal fluctuations get smoothed and diluted by the sheer magnitude of the Earth's heating and ventilation systems. A few times in Earth's past, however, these systems seem to have broken down catastrophically. In each case, an unusual combination of circumstances interfered with ocean–atmosphere mixing, and the Earth climate became wildly unstable.

Many are cold but few are frozen

One such episode occurred between about 750 and 600 million years ago, very close to the end of the long Precambrian eon and just before the appearance of modern lineages of animals. This is the stranger-than-science-fiction Snowball Earth interval.[16] In a handful of regions around the world, most of them distinctly inhospitable (arctic Norway, the Australian outback, Death Valley, the deserts of Namibia), sedimentary sequences tell us that something was amiss with Earth's climate in latest Proterozoic time. First, at nearly every site where rocks of this age are exposed, one finds **diamictites**, the disorganized sedimentary rocks with two distinct populations of grain sizes, reflecting deposition by glacial ice or icebergs laden with rocky debris. Diamictites are always interesting, but they are found elsewhere in the geologic record. What makes these particular glacial deposits remarkable is that their paleomagnetic signatures indicate deposition at very low latitudes. In other words, there was ice not only at the poles and high elevations but *at sea level at or near the equator.*

The presence of ice at these low latitudes is disturbing, but odder still are two other types of rocks that occur just above the diamictites at many of the sites: carbonate rocks (limestone and dolomite), which typically form in warm, tropical seas, and **banded iron formations**, a chemically precipitated rock, which like komatiite had long been "extinct" at the time that the diamictites were deposited.

Most of the world's iron-rich sedimentary rocks, or banded iron formations, were laid down in early Proterozoic time (about 2.5 to 2.0 billion years ago), at a time of critical change in the Earth's atmosphere. Before about 2 billion years ago, there was very little free oxygen in the atmosphere, and large quantities of iron could exist in dissolved form in the oceans, just as calcium and sodium occur as ions in seawater today. When iron becomes oxidized, however, it is highly insoluble and will quickly precipitate out of water in solid form as a layer of iron-rich sediment. The immensely thick early Proterozoic banded iron formations, including those in northern Minnesota and the Upper Peninsula of Michigan (which were the foundation of the U.S. steel industry throughout the twentieth century), are thought to record the transition from conditions of **reduction** to conditions of **oxidation** as photosynthesizing microorganisms began to control the Earth's atmosphere. Once all the previously dissolved iron had dropped out of the oceans, iron formations stopped being deposited. At the time that the low-latitude diamictites were accumulating, more than a billion years had passed since the last major iron formations were laid down. So the sudden reappearance of this supposedly discontinued rock type requires explanation.

But the Snowball Earth hypothesis seeks to account not only for the diamictites and iron formations but also their strange sedimentary bedfellows, the "cap" carbonates, which consistently top the glacial deposits wherever they are found. According to the hypothesis, these limestones are records of a sweltering global heat wave that followed an extreme ice age.[17] If the theory's principal proponents weren't well-respected geologists, this decidedly nonuniformitarian scenario would be a tough sell. But its authors (who include Joe Kirschvink of Cal Tech, Paul Hoffman of the Canadian Geological Survey, and Dan Shrag of Harvard University) are no slouches. And the fact that the story accounts for so many vexing irregularities of the latest Proterozoic rock record has made a surprising number of geologists willing to take the tale seriously.

The hard version of the Snowball Earth story holds that the entire globe, including the oceans, was covered with ice. This arrangement would have prevented the atmosphere and oceans from interacting via evaporation, precipitation, and wind-driven turbulence. Under normal circumstances, the upper part of the ocean and the atmosphere are a well-mixed system; the shallow oceans are therefore rich in oxygen and support a thriving community of organisms. The reappearance of banded iron formations in the late Proterozoic sequences suggests that the oceans somehow became oxygen-depleted and able for a time to hold iron in solution. The only way to deplete the ocean of oxygen would be to seal the oceans off from the atmosphere, and a layer of sea ice would do the job nicely. The iron-rich strata themselves would have been deposited once the ice had melted and the ocean–atmosphere oxygen dialogue resumed.

But the proposal of an Earth encased completely in ice is problematic. The idea was first explored by a Russian climate modeler, Mikhail Budyko, who in the 1960s had learned of the extensive late Proterozoic diamictites on the arctic archipelago of Svalbard, a territory to which both Norway and Russia lay historical claim. (I myself feel a little proprietary about the Svalbard diamictites because I studied these rocks as part of my Ph.D. dissertation work.)

Budyko developed a simple quantitative model of the increasing global albedo (reflectivity) caused by the unchecked growth of ice caps. His model showed that ice sheet growth involved strong positive feedback—the more snow and ice, the higher the albedo, the colder the temperatures, the more snow. Once glaciers reached tropical latitudes, there was no turning back; Earth would be locked into an irreversible deep freeze.[18] Budyko therefore rejected the speculation that the diamictites of Svalbard, which are both underlain and overlain by tropical-looking carbonates, could truly have represented low-latitude deposits.

Budyko's model was still on skeptics' minds when, in the early 1990s, Joe Kirschvink introduced new paleomagnetic evidence that some of the late Proterozoic diamictites had indeed formed at near-equatorial latitudes. Kirschvink himself briefly speculated that the Earth's axis had flopped over so that it lay on its side, with its erstwhile poles exposed to maximum sunlight and the equatorial regions suddenly cool and shady.[19] This idea has been quietly forgotten, but the problem remains: How could the Earth ever extricate itself from an ultra–ice age?

The best way to warm a hypothermic planet is with a blanket of greenhouse gases. Paul Hoffman's and Dan Schrag's principal contribution to the snowball story was to point out that even if the hydrosphere, the atmosphere, and much of the biosphere were locked into a cryogenic slumber, volcanoes would have remained awake. With a far less active biosphere to take up the exhalations of volcanoes, carbon dioxide levels may have climbed to extreme values (Hoffman has suggested carbon dioxide concentrations as high as 120,000 parts per million, 300 times higher than even the fossil-fuel-enhanced levels of today). This carbon dioxide buildup could have taken as long as 10 million years, but finally the Earth would have shaken the chill—only to succumb to a supergreenhouse fever. The huge carbon dioxide load in the atmosphere would have created fierce acid rain (under normal circumstances, even unpolluted rainwater is slightly acidic because dissolved carbon dioxide in it creates weak carbonic acid). Acid rain would have accelerated the rates of chemical weathering of rocks on the newly deglaciated continents, sending large amounts of calcium and other salts into the ocean. The calcium carbonate rocks that cap the diamictite layers may record the rapid precipitation of these dissolved ions as they encountered seawater that was bubbling with carbon dioxide.

Skeptics question whether even the extreme carbon dioxide concentrations that Hoffman and Schrag propose would have been sufficient to pull the Earth out of its icy blues.[20] Paleobiologists also question whether the biosphere could have survived a 10-million-year ice age in which even the ocean became nearly barren of life. (Hoffman and Schrag suggest that submarine volcanoes would have been the arks that kept microbial life from being extinguished during the long global winter.)

But there is another, quicker way to pump greenhouse gases into the atmosphere: the destabilization of vast stores of methane. Put crudely, planetary burping. This dyspeptic scenario draws on the growing evidence that immense volumes of biogenic methane occur on the seafloor in the form of **gas hydrates,** or icy mixtures of water and natural gas.[21] Methane and other wastes produced by marine microorganisms can take the form of ice under a narrow range of pressure and temperature conditions within sediments of the seafloor. But they will rapidly volatilize (break up into gaseous form) when these conditions change, as a result of, for example, changes in ocean temperature or sea level. The rapid release of large stores of methane would have a tremendous effect on climate because methane is more than twenty times more effective as a greenhouse gas than carbon dioxide. So it would take far less methane than carbon dioxide to melt the snowball, and methane from gas hydrates could potentially have been released over much shorter times (perhaps centuries) than it would have taken volcanoes to cough up enough carbon dioxide (millions of years).

Again, the cap carbonates are the rocks we should consult for information about the great thaw. Not all carbon atoms are alike. The most common form is carbon-12 (^{12}C), with six protons and six neutrons, but there is also a heavier variety, ^{13}C, with seven neutrons. Both ^{12}C and ^{13}C, in contrast to their radioactive cousin, ^{14}C, are stable and do not break down over time (after about 60,000 years, any ^{14}C in rocks and organic matter will have decayed away). Although the difference in atomic weight of ^{12}C and ^{13}C is small, it is just enough that photosynthesizing plants and phytoplankton will "avoid" the heavier ^{13}C since it takes just a little more energy to extract the heavier form of carbon dioxide from the atmosphere. So biologically fixed carbon tends to be isotopically light, with relatively more ^{12}C than the "raw," unsorted carbon coming out of volcanoes.

The carbon in the cap carbonates from the Snowball Earth interval is in fact significantly enriched in ^{12}C, which suggests that it came largely from a biological source rather than from accumulated volcanic emissions. The evidence for a biological source for the carbon seems more consistent with the gas hydrate scenario, although Hoffman and Schrags argue that the isotopic signature in the cap carbonates could also be explained by the absence of significant photosynthesis in the oceans for millions of years. With no photosynthesis occurring, seawater would accumulate unusually high amounts of ^{12}C that otherwise would have been taken up by organisms. The debate hinges on the exact timing and duration of the carbon isotope variations, and so far, the rocks haven't provided enough information for geological detectives to distinguish events that may have happened over thousands of years from those that happened over millions.

While there is continuing debate about whether the latest Proterozoic Earth was truly snowbound or just slushy, and whether methane belches or volcanoes were responsible for the thaw, this episode was clearly one of the greatest climatic crises the planet has ever experienced. In fact, the deposits on Svalbard and several other classic sites indicate that there was not just one but as many as four freeze–thaw cycles over a period of about 150 million years. Why did these ice ages happen in the first place? There is no reason to think that the Sun went dim for a few hundred million years. And the orbital variations that seem at least partly responsible for the more recent Pleistocene glaciations happen on timescales of tens to hundreds of thousands of years. Greenhouse gases must therefore be implicated, but what could cause the Earth's carbon spending habits to change over such long periods?

A hundred million years is the approximate life span of an ocean basin or a supercontinent, and in latest Proterozoic time, almost all of Earth's continental crust was jammed together in a vast landmass posthumously called **Rodinia** (this landmass is a generation older than the more famous supercontinent **Pangaea**). Supercontinents come and supercontinents go, but Rodinia was different in that it straddled the equator, so that most of it lay within tropical latitudes. This location is relevant to the planet's long-term climate control system because one of the most important ways that Earth withdraws carbon from the atmosphere is through the weathering of silicate continental crust. As in the post-snowball scenario, but usually at a far gentler pace, atmospheric carbon dioxide combines with rain, forming carbonic acid, which dissolves calcium and other ions from rocks and washes them out to sea, where they are precipitated, often with biological assistance, as carbonate rocks. This ingenious system is the most important difference between Earth and its hopelessly overheated sister planet, Venus, which has never learned to put away its carbon.

But in latest Proterozoic time, Earth was apparently withdrawing carbon from the atmosphere too quickly, and the low-latitude positioning of Rodinia may have been the reason. In ordinary ice ages, like the Pleistocene, as ice advances out of the polar regions, it begins to cover up landmasses, thereby slowing the weathering-related drawdown of atmospheric carbon dioxide. This in turn warms the climate a bit and keeps the ice from continuing to advance (Mikhail Budyko's model of ice sheet albedo did not include this negative feedback). But if much of the Earth's continental crust lies at low latitudes, as it did in the late Proterozoic, rock weathering will continue unchecked until glaciers reach the tropics. When mantle convection finally led to the breakup of Rodinia and continental crust was dispersed again to higher latitudes, the normal conversations between rock, water, and air could resume. But as we will see in the next chapter, life on Earth had been irrevocably changed.

Permian permutations

For the next several hundred millions of years, during the Paleozoic era, Earth was on an even keel. Increasingly complex lifeforms thrived in the oceans, and by the Silurian period (about 425 million years ago), a small group of pioneering plants and animals had begun to colonize the previously uninhabited continents. By the Carboniferous period (300 million years ago), mammal-like reptiles walked the floors of lush forests. Other than a couple of ice ages and large meteorite impacts (both of which are implicated in mass extinctions in the Ordovician and Devonian periods), the Paleozoic was a time of comparative stability, at least relative to the mercurial conditions of Snowball Earth time. Then, late in the Permian period, about 250 million years ago, something went terribly wrong.[22]

In a period of perhaps one million years, the global ecosystem collapsed, and an estimated 90 percent of all species went extinct. This is the largest mass extinction event in Earth's history, far more severe than the more infamous Cretaceous–Tertiary (K–T) event that killed the dinosaurs (the extinction rate in that event was about 65 percent). In contrast to the K-T catastrophe, which most geologists now believe was related to the impact of a large meteorite off the coast of the Yucatan, the Permian extinction apparently had an internal, earthly cause. As in the case of the Snowball Earth interval, the planet's "metabolism" seems to have gone haywire. This event remains one of the most frightening and least understood intervals in Earth's past. Eerie parallels between the late Permian and modern times make it important to understand what happened in this close brush with death.

In the millions of years just before the biosphere began to unravel, a new supercontinent, Pangaea, had been built, reuniting the dispersed fragments of Rodinia in a new configuration. In early Permian time, a moderate ice age occurred, but because Pangaea, unlike Rodinia, spanned high and low latitudes, the glaciers did not advance very far out of the polar regions. Global sea level dropped for a time during the glacial period, reducing the area of the continental shelves. This loss of shallow-water habitat may have placed some stress on marine ecosystems, which preferentially occupy these light-suffused, nutrient-rich areas. But this moderate glaciation event and the concomitant sea level change occurred millions of years too early to be solely responsible for the great extinction event, and they cannot account for its brevity and severity.

High-resolution uranium–lead zircon dates from end-Permian strata in Texas and southern China suggest that the main pulse of extinction was quick and brutal, perhaps as short as 900,000 years (from 252.3 to 251.4 million years ago).[23] At this same time, bad things were happening to the Earth's air and water. Volcanic fissures in Siberia were spewing out basaltic lava, as well as carbon dioxide, sulfur dioxide, chlorine, and fluorine at nearly unprecedented rates, in one of the largest subaerial volcanic events in Earth's history. The rocks formed in this flood basalt event are called the Siberian traps. (*Traps*, or *traprock* is a quaint old term for basalt, derived from a Norse word for stairs [*trappe*], a reference to the steplike shapes formed by eroded basalt layers.) The Siberian traps have always been considered a world-class flood basalt sequence, ranking in the top four with those in the Columbia River Gorge (ca. 15 million years old), the Midcontinent Rift of the Lake Superior region (1.1 billion years old), and the Deccan region of India (65 million years old—curiously, the precise time of the K–T extinction event). But recent drilling has shown that the Siberian traps may have been far more extensive than previously thought, covering an area of some 800,000 square miles, more than the twice the size of Europe.[24]

There were other changes in the air. In south China and Svalbard (a small place with a rich geologic legacy), which even then were on opposite sides of the globe, marine sediments spanning the transition from the latest Permian into the beginning of the postapocalyptic Triassic period suggest that ocean life was suffocating. Red and green Permian mudstones teeming with fossils give way to black, barren, sulfurous, carbonaceous sediment similar to the reeking organic slime that collects at the bottom of the modern Black Sea. (Geologists use the term *euxinic* to describe such dark black shales; the word comes from the ancient Latin name for the Black Sea: Pontus Euxinus.) The mudstones' change in color from rosy red to funerary black, the partly decomposed organic matter, and the presence of iron sulfide minerals like pyrite, all indicate that in end-Permian time, and well into the early Triassic, oxygen had completely disappeared from the previously thriving seafloor ecosystems.[25] Apparently there were not even scavengers around to take advantage of all the rotting matter.

The anoxic (oxygen-depleted) dead zone beneath the surface waters in the modern Black Sea (and in a growing number of other areas, including the Gulf of Mexico and Chesapeake Bay) has formed because large amounts of decomposing organic matter have sucked the oxygen out of the deeper reaches. The sea is also strongly stratified, with dense, salty bottom water trapped beneath fresher, lighter water. This makes convective overturn, which would freshen and aerate the bottom waters, impossible. For the oceans to go anoxic, as they probably did at the end of the Permian, means that they went similarly stagnant and ceased to communicate with the atmosphere. Could the Siberian volcanoes really have caused all this?

During the peak of the eruptions, the large volumes of the volcanic gases carbon dioxide, sulfur dioxide, chlorine, and fluorine would have turned the rain into a burning mix of carbonic, sulfuric, hydrochloric, and hydrofluoric **acids** (it gives one pause to realize that these volcanic emissions are not terribly different from the output of modern smokestacks). Terrestrial ecosystems might have been hit hard, but the calcium-rich oceans, like a giant antacid tablet, could probably have buffered the acid precipitation. If nothing else had happened, the debilitated land-based ecosystems might have recovered. But another environmental blow, and possibly several others, led to a catastrophe of global proportions.

Carbon and oxygen isotopes from end-Permian rocks tell the horrific tale. We can use the oxygen isotopes from calcite ($CaCO_3$) in limestone to infer the temperature of the seawater from which the mineral was precipitated much as we use the oxygen isotopes in glacial ice (H_2O) to reconstruct air temperatures at the time that the original snow fell (Chapter 3). Oxygen isotope ratios from latest Permian limestones show that global temperatures rose by as much as 11°F over a comparatively short time. The limits of age dating methods make it impossible to know exactly how rapidly this temperature increase occurred, but a temperature jump of that magnitude (equivalent to the difference between modern temperatures and those at the close of the last Ice Age) would have placed considerable additional stress on an already weakened global ecosystem. The paleotemperature data also suggest a degree of greenhouse insulation greater than that expected from the Siberian volcanoes, since the warming caused by the carbon dioxide emissions would have been partly offset by the cooling caused by the reflectivity of sulfur particles in the upper atmosphere. The late Permian carbon isotope record in both marine limestones and terrestrial **paleosols** (clay-rich sedimentary rocks interpreted as ancient soils) suggests that again an oceanic belch of methane may have pushed the climate system over the edge.

Like the cap carbonates of the Snowball Earth interval, the Permian soil deposits contain exceptionally light carbon.[26] Decaying organic matter, perhaps in the form of gas hydrates, seems to be the only reservoir of carbon that could account for the abrupt shift to such isotopically light values. Asphyxiating quantities of carbon dioxide and methane may have been released over a period of just weeks. (A sudden, deadly release of natural carbon dioxide from Lake Nyos in Cameroon in 1986, which killed nearly two thousand people, is often cited as a small-scale analog to the Permian calamity.) So whatever devastation the Siberian volcanism caused, the belches of organic gas would only have made the situation worse. Massive die-offs of both terrestrial and marine life would have turned the seas anoxic with decomposing organic matter, perhaps triggering additional releases of suffocating gases. Ecosystems would have been hit broadside, with serious damage at every trophic level, from primary producers to top predators.[27] The survivor profile is sobering; nothing larger than a house cat walked across the Permian–Triassic boundary, and the terrestrial ecosystem evidently did not fully recover for as much as 5 million years.[28]

This nightmarish combination of events would seem quite enough to account for the Permian cataclysm, but some scientists are still looking for extraterrestrial causes. A group led by Luann Becker of the University of California at Santa Barbara has recently suggested that they have located an impact crater of the right age on the continental shelf of northwestern Australia.[29] If a large meteorite did slam into the Earth at this time, it would have been just another in a series of cruel blows. Douglas Erwin has compared geologists' difficulty in assigning culpability for the Permian extinction to the job faced by detective Hercules Poirot in *Murder on the Orient Express*. In the end, Poirot realizes that the victim (himself a serial murderer) was stabbed in sequence by twelve other passengers, each with a reason for seeking revenge. Is it more comforting to think that the Permian catastrophe was caused by the unlikely convergence of a series of events or by a single nefarious villain? In a time when anthropogenic emissions of sulfur and chlorine match or exceed volcanic releases, when human carbon dioxide production outstrips natural rates by a factor of ten, and when growing areas of the world's oceans are becoming dead zones as a result of sewage and fertilizer runoff, I'm not sure. More recent records of climate instability are equally sobering.

The day after the Pleistocene

On the Wisconsin shore of Lake Michigan, a mile east of a one-horse town called Two Creeks, local teenagers have long stoked campfires on the beach with wood from a layer of buried stumps and logs exposed about halfway up the wave-cut slopes. If you bothered to count the rings, you would find that some of the bigger trees were as old as 180 years when they were buried in the sand and clay. In addition to these large pieces, smaller branches and twigs as well as pine and spruce cones and needles protrude from the layer. The stuff looks very much like the soft, fragrant blanket of evergreen litter on the floor of a pine woods in northern Wisconsin. But if you took the time to examine the sediment above and below the woody layer, you would see with a shock that the "forest floor" is sandwiched between two bouldery clay strata that are unmistakable glacial deposits. This means that after the retreat of an ice sheet, a forest took root and thrived for nearly two centuries, then was abruptly buried (thereby preserved from decomposition) and overridden again by glacial ice.

When this wood was sampled for carbon dating in the early 1950s (it was in fact among the first materials ever dated using the ^{14}C method), it yielded the astonishing age of 11,700 years.[30] Unraveling the story of the Two Creeks buried forest became the

Ph.D. project for a student at Columbia University, Wallace Broecker, who would become one of the leading thinkers in the emerging discipline of paleoclimatology. Broecker recognized that the Two Creeks layer clearly documented a rapid return to full glacial conditions late in the Ice Age, after a period when the glaciers had been melting back for at least 1,000 years. Broecker and others began to gather evidence that the cold snap at Two Creeks was a global phenomenon, which today is known in the paleoclimate community as the **Younger Dryas**. Just before the Earth emerged from the grip of the last Ice Age, things got very cold again for a millennium or so. What could have caused such an abrupt change in climate?

We have seen how volcanic exhalations and oceanic belches can abruptly raise global temperatures by rapidly increasing the greenhouse gas content of the atmosphere, but there is no reciprocal process by which carbon dioxide and methane can be rapidly drawn out of the atmosphere. The timescale is far too short to invoke silicate weathering as in the Snowball Earth scenario. So Broecker had to look for other mechanisms by which Earth's heat budget could change over a period of a century or less. The answer was the movement of water in the oceans.

Earth's ocean currents are the planet's heat distribution system, equalizing disparities in the amount of solar radiation received at different latitudes. Large currents like the Gulf Stream transport warm water from the tropics toward the polar regions. Without this imported heat, Britain and northern Europe, which lie at the same latitudes as do Alaska and northern Canada, would be far less hospitable. Growing seasons would be so short that agriculture would not be possible. As the Gulf Stream water moves northward and lends its heat to the surrounding lands, the water not only cools but also becomes saltier, owing to repeated cycles of evaporation during its journey. Ultimately, in the Norwegian–Greenland Sea, the water becomes so dense that it sinks again. From this point, it begins an epic journey to the south, snaking deep below the surface as bottom water, eventually reaching the Indian Ocean, where, warmed again, the water may rise and meet the atmosphere once more.

This **thermohaline ocean circulation** is driven by convection. Like the much-longer-term convection in Earth's mantle, thermohaline ("heat-salt") circulation requires a critical balance of variables. If the salinity of the water in the North Atlantic did not reach the critical value for sinking, the entire conveyor could stop, just as a single stalled car on a highway can back up traffic for miles. Wally Broecker and his colleagues recognized that a rapid drop in the salinity of the North Atlantic could explain the sudden chill recorded by the Two Creeks layer.[31] Late in the last Ice Age, as the continental glaciers covering northern North America and Europe were rapidly disintegrating, huge amounts of fresh meltwater would have flooded the North Atlantic. Vast glacial lakes like Lake Agassiz in North Dakota and Manitoba (which was larger than all the modern Great Lakes combined) may have spilled out catastrophically into Hudson Bay and the Saint Lawrence Seaway when their previously ice-dammed outlets were opened.[32]

The fresh meltwater would have diluted the waters coming up from the south to the point where the resulting, less saline water had no inclination to sink. Traffic on the oceanic highway—previously traveling at a rate of hundreds of sverdrups (the equivalent of a thousand Mississippis)—would have been jammed for decades or longer as more and more meltwater came streaming off the land. Cut off from its source of imported sunshine, the North Atlantic and surrounding land areas would have become colder and colder, triggering a brief return to glacial conditions. This in turn would have allowed the Gulf Stream to reestablish itself, warming the north again, and finally pulling the Earth out of the Ice Age.

Broecker's ideas, as intoned by Dennis Quaid, made it to Hollywood in the 2004 thriller *The Day After Tomorrow*. The premise of the film is that human greenhouse emissions have triggered rapid melting of the ice masses, causing the sudden freshening of the North Atlantic and the concomitant shutdown of the thermohaline conveyor. Extreme weather ensues over a period of about a week, and when the snow settles, the Statue of Liberty is up to her armpits in ice. The film's timescale is absurdly compressed, but the underlying science is sound. The trees at Two Creeks tell us that we should start imagining climate change in our own lifetimes.

ONLY CONNECT

Understanding the rare times when Earth apparently lost its sense of balance makes the planet's general equanimity seem all the more remarkable. Like a bicycle rider, Earth is most stable when its wheels are turning. Stagnation is death; recycling is the overriding rule. The system is robust as long as shallow and deep waters roll over each other, oceans and atmosphere mix freely, and rock and water interact at all latitudes and depths, even deep into the planet's interior. In spite of its inherent risks, connectivity can always provide greater stability than isolation.

Comparing Earth with its childhood peers, Mars and Venus, we see that even if a planet has all the raw materials, it will not necessarily develop a sustainable tectonic system, consistent climate, or long-lived biosphere. The materials must first be organized, and then they must "learn" to interact at all timescales in a manner that balances the processes of sorting and concentration (volcanism, evaporation, photosynthesis) with those of mixing and dispersal (subduction, weathering, decomposition). Both mixing and sorting have the capacity to be constructive or destructive;

neither is the villain or the hero of the story. Unchecked dissipation, like extreme segregation, can cripple a healthy system. But coupled in an intimate give and take, mixing and sorting—maelstroms and neatniks—form a potent creative team that has allowed the planet to maintain its vigor for 4 billion years.

5

INNOVATION AND
CONSERVATION

Too much safety yields only danger in the long run.

— ALDO LEOPOLD

Because we don't think about future generations, they will never forget us.

— HENRIK TIKKANEN

YOU SAY YOU WANT A REVOLUTION

IN EVERY FAMILY, in every generation in every society, tensions between the old and the new define relationships, popular culture, and politics. Caretakers (also known variously as conservatives, fuddy-duddies, and keepers of the faith) are pitted against innovators (i.e., progressives, troublemakers, visionaries). How do we decide what to keep and what to jettison? When is it time to try something altogether novel, and when is it best to stay with the tried and true? Is there in fact anything new under the

Sun? And if so, do radical innovations stand a chance of surviving in a world of entrenched habits?

In Earth's past, the forces of conservation and innovation have been yoked together, alternately leading and lagging, but always keeping each other in check. In the biosphere, conservation is favored in times of stability, innovation in times of stress. Some of the greatest evolutionary advances have followed the most devastating environmental crises in the Earth's past. But are the organisms and survival strategies that emerged from such times truly fitter or better in an absolute sense? To what extent have a few early, ad hoc designs precluded further experimentation? The rock record has both sobering and inspiring tales for those who would change the world.

THE PARADOX OF OXYGEN

The worldwide deposition of banded iron formations from about 2.5 to 2.0 billion years ago (Chapter 4) records one of the most profound changes in the Earth's surface environment—the transition from an atmosphere dominated by volcanic carbon dioxide and water vapor to one in which free oxygen was a major constituent. Before this, iron, which is very soluble in reduced form, was present in large quantities in seawater. But by early Proterozoic time, photosynthesizing **cyanobacteria** (blue-green algae like the ones that create slimy scums on ponds) had been spewing out oxygen for at least a billion years, and there were changes in the air. The oceans had at last become oxygenated, and all the dissolved iron, now insoluble, was precipitated in massive, rusty layers over a geologically short interval. Only when the oceans were clear of oxygen-greedy iron could the gas begin to accumulate to significant levels in the atmosphere.

This stage of the geochemical regime change is also noted in the sedimentary record. Just as banded iron formations vanish, a brave new type of strata, **red beds** (an appropriately revolutionary name) appear. These represent terrestrial (land) deposits, laid down in rivers or in fans of sediment at the feet of ancient mountains. The beds' distinguishing characteristic is their russet color, which indicates that the weathered surfaces of iron-bearing minerals in these sediments were oxidized at the time of deposition. Red beds tell us that the oxygen revolution had occurred by about 1.8 billion years ago.

The revolution had been brought about by the photosynthesizers, but it created a world that many of the instigators themselves found inhospitable. For most denizens of the diverse but almost exclusively single-celled early biosphere, oxygen was toxic. Oxygen had been only a local waste disposal problem while there was still plenty of reactive iron to soak it up, but once this gas became ubiquitous in the atmosphere, it was a global environmental crisis. A radical response was needed, and the emergency plan that happened to work defined the subsequent evolution of the biosphere.

As early as 2.5 billion years ago, well before oxygen was omnipresent, some entrepreneurial single-celled organisms, including a group called the mitochondria, were beginning to take advantage of the oxygen "exhaust" emitted by photosynthesizers. When oxygen displaced carbon dioxide in the global atmospheric regime, these organisms were ready for the coup d'état. Others were not, however, and their options for survival were limited. Organisms at this time were simple in design—**prokaryotes** lacking cell nuclei or other specialized parts (organelles) to which various types of domestic chores could be delegated. One choice for oxygen-phobic organisms was to flee to enclaves where the chemical milieu would allow them to live as they always had.

Today, the descendants of some of these anaerobic organisms thrive in environments like swamps, the depths of stagnant water bodies, and the stomachs of ruminants like cattle, where the local environment more closely resembles Earth's early atmosphere.

Alternatively, oxygen-unprepared organisms could remain at Earth's surface by forging alliances with those that had planned for the new world order. In a brilliant strategic merger, some early anaerobes assimilated mitochondria, which allowed the anaerobes to use oxygen as a metabolic fuel. The tiny mitochondria, in turn, received room and board (shelter and nutrients) from their hosts. Every cell in our bodies, and in every plant and animal on Earth, records this symbiotic union. Mitochondria are essential components of our cellular machinery—the respiration power stations where energy is extracted from glucose and made available for essential metabolic functions. (The word *mitochrondria* means "thread grains" and shares a Greek root with *chrondrules,* those even more ancient grains of stardust that may have seeded Earth with the raw materials for life.) Strangely, mitochondria have their own DNA, which is entirely separate from the host organism's own DNA, housed in the cell nucleus. (Because mitochondrial DNA is not mixed through the genetic exchanges that occur in sexual reproduction—it is inherited intact from one's mother—it conserves information about lineages. It has been used to infer when and where "Eve," the last common female ancestor of all modern humans, lived.)

Microbiologist Lynn Margulis was among the first to argue that the DNA in mitochondria is evidence that they were once independent organisms that boarded with our ancient one-celled ancestors and decided to stay.[1] Subsequent work by Margulis and others has shown that this kind of **endosymbiosis** was a common evolutionary tactic in the biosphere. Chloroplasts, the photosynthesizing apparatuses within plant cells, probably originated as independent cyanobacteria that later joined forces with early **eukaryotes** to form a new kingdom. The cell nucleus itself, which is the defining characteristic of the eukaryotes, may have originated through a merger of the same type, at about the same time as the mitochondrial union. This inference is supported by the observation that virtually all eukaryotic organisms (including humans) are oxygen lovers whose cells contain mitochondria. Moreover, the first microfossils with clear cell nuclei appear in the fossil record at the time of the oxygen revolution. Fossils of the coiled alga *Grypania spiralis*—perhaps the progenitor of all modern eukaryotic organisms—have been reported from a 2.1-billion-year-old banded iron formation deep in a mine in the Upper Peninsula of Michigan, although there is some disagreement about how to interpret those tiny ringlet-like forms.[2]

The altered atmosphere affected the biosphere in still other ways. A fortuitous by-product of the new oxygen-rich skies was the stratospheric **ozone** layer, produced by the interaction of free diatomic oxygen (O_2) with ultraviolet (UV) radiation from the Sun. In the uppermost atmosphere, short-wavelength ultraviolet light dissects some of the O_2 molecules into single oxygen atoms, which then react with intact O_2 to form ozone, O_3. The ozone, in turn, is itself continuously bombarded by ultraviolet light, and when struck by slightly longer wavelengths (known as UV-B), it breaks up again into O and O_2. In this way ozone acts to absorb UV-B radiation, which is particularly harmful to living organisms, because it damages DNA. Under natural conditions, the rates of ozone creation and destruction are about balanced, so the amount of stratospheric ozone remains approximately constant.

Life on Earth had emerged and initially evolved without the benefit of the ozone shield, and the hardy pioneers that first colonized the planet, especially the cyanobacteria, had to excel at repairing damaged DNA and otherwise mitigating the injurious effects of UV irradiation. In a pre-ozone world, selective pressures for high-fidelity DNA replication would have been great, at least for organisms living in shallow waters (even today cyanobacteria preserve some of the most ancient genetic sequences of any group of organisms).[3] In early Proterozoic time, however, the fierceness of genetically damaging UV radiation began to wane just when new geochemical rules were being established, and the capacity for genetic innovation would have become favored.

The easing of the radioactive assault and the growing benefits of genetic innovation could explain the timing of the invention of sexual reproduction. The rock record is demurely quiet about this defining moment in the evolution of life, but paleobiologists surmise that it occurred sometime around the middle Proterozoic.[4] The origin of sex is a long-standing puzzle in evolutionary biology. Slower and riskier than one-parent reproduction, sexual reproduction is nonetheless the rule for most multicellular, and many unicellular, organisms. Evolutionary biologist John Maynard Smith quantified this paradox with his famous model of a population of sexually reproducing organisms in which a mutation arises that allows parthenogenic reproduction (cloning) by females (as is the case for some aphid species). Assuming that neither the number nor the viability of the offspring produced by a given female depend on her mode of reproduction, it is easily shown that parthenogenic females will comprise a growing proportion of the population with each generation, and that, ultimately, sexually reproducing individuals will disappear.[5] The persistence of sex seems even more puzzling considering that parthenogenetically reproducing individuals typically produce more offspring than do sexual ones, contrary to the first assumption in Maynard Smith's model. Sex must therefore, in some circumstances, enhance the probability of offspring survival enough to counteract the inherent numeric advantages of uniparental reproduction.

While there is continuing uncertainty about the evolutionary advantages of sexual reproduction, most models and experimental studies support the view that sex enhances the efficiency of adaptation by natural selection.[6] Sexual reproduction allows the reshuffling of genetic material and reduces the incidence of a phenomenon called *linkage disequilibrium*, in which functionally unrelated genes become locked together on the genome. Such linkage means that a disadvantageous trait may hitchhike with an advantageous one, and this has a braking effect on evolutionary rates.[7] In contrast, sex allows for genetic experimentation with a limited number of uninvited passengers, enabling organisms to develop finely tuned adaptations to changing environments.[8] It seems plausible that sex arose at a time when new environmental conditions began to favor its capacity for evolutionary innovation.[9] At present, however, paleontological and molecular data are insufficient to permit a conclusive determination of the geological moment when sex became the preference of the eukaryotes.

The lessons of the oxygen revolution would seem to be that it pays to anticipate change, as the mitochondria did, but when change is afoot, it is even more important to forge strong ties with able allies. In times of crisis, the most successful and versatile innovations can come from symbiotic cooperation, creative synthesis of acquired skills, and the free exchange of information.[10]

COMING OUT OF THE COLD

After the great evolutionary leaps of the early Proterozoic, the fossil record of the second part of the Proterozoic is comparatively uneventful. Eukaryotes became well established and diverse, coexisting peaceably alongside prokaryotes that kept to their own low-oxygen haunts. **Stromatolites**—finely laminated deposits left by mats of cyanobacteria and other microbes—are abundant in the stratigraphic record of this time. The most notable invention of this time is multicellularity in eukaryotes—not merely the concatenation of single cells into strings of beads (this is seen even in the very earliest fossils, from Archean time) but a new, more complex body system in which different cells perform different functions. So far, no one has been able to find the exact page in the rock record when multicellular animals first emerged, but by middle Proterozoic time (ca. 1.2 billion years ago), something— probably a primitive worm—left "scribbles" on sandstone beds in southwestern Australia.[11] If these are crawling tracks, they could only have been produced by an organism sophisticated enough to perambulate across the seafloor. More conclusive evidence of multicellularity comes from fossils thought to be of a type of red algae with differentiated cells, preserved in strata on Somerset Island (arctic Canada) at about the same time as the tracks of the Australian crawler.[12]

Even if multicellular animals had emerged by mid- to late-Proterozoic time, they appear to have had no huge evolutionary advantage over their simpler predecessors. The organisms that dominate the later Proterozoic fossil record are single-celled structures called *acritarchs*, which probably represent a stage in the life cycle of simpler varieties of algae. At up to 0.1 inch in length, they were giants in their time. They were hugely diverse and successful, dominating the oceans for hundreds of millions of years. Acritarchs reached their zenith around 850 million years ago, and then, as the planet slipped into the grip of the Snowball Earth glaciations, they were very nearly wiped out.

When Earth finally recovered from its shivers around 570 million years ago, the environmental decks had evidently been cleared. Niches that had been the strongholds of acritarchs and other organisms were left vacant after the harsh ice ages. For perhaps the first time since the rise of oxygen, the opportunities for entrepreneurs were limitless. The first homesteaders to stake their claims in this frontier world are an enigmatic group of fossil organisms called the **Ediacaran biota**, after the region of their discovery in southern Australia. Also known as Vendozoans or Vendobionts (*Vendian* is an older name for the Snowball Earth interval, introduced by Russian geologists in the 1950s), these alien-looking organisms have been variously interpreted as lichenlike symbionts of fungi and algae, ancestors of modern arthropods and jellyfish, and as a failed line of early animals with no living descendants.[13] This last, rather disturbing, interpretation has gained acceptance as fossil beds in Namibia, Sweden, England, and Newfoundland have yielded spectacular new insights into Ediacaran anatomies and lifestyles.

All of the Ediacaran organisms seem to have been soft-bodied. Many look puffy and quilted, like small versions of old-fashioned air mattresses. One group resembles modern sea pens, with clearly preserved holdfasts that must have anchored them to the seafloor. Some of these were spectacularly, and unprecedently, large: One frondlike fossil called *Charnia* grew as tall as five feet. Other Ediacarans look like worms but (surprisingly) lack evidence for any kind of digestive tract. They apparently nourished themselves either by drawing nutrients directly from seawater or by hosting symbiont photosynthesizing microorganisms within their bodies. The fossils at the ominously named Mistaken Point locality in Newfoundland are exquisitely preserved in volcanic ash and record a sophisticated ecosystem in which different types of organisms lived at different water depths.[14]

The Mistaken Point ash layers also yield zircon crystals whose ages reveal that the earliest Ediacaran organisms appeared more than 570 million years ago, almost immediately after the final Snowball Earth glaciation.[15] For the next 30 million years, the quirky Ediacarans defined an altogether novel kind of biosphere on a planet that had nearly frozen to death. Then, about 540 million years ago, these airy, pliant organisms disappeared. Somewhere in the world, a radical and ruthless new evolutionary strategy had emerged: predation. The Ediacarans didn't have a chance.

SWIMMING WITH THE (NOT-YET-EVOLVED) SHARKS

The renowned Burgess Shale fossil beds of British Columbia (Chapter 1) provide the most complete picture of the brutal world that supplanted the Ediacaran Eden.[16] But before diving into the perilous middle Cambrian seas, we need to peer into the equally treacherous waters of paleontology and evolutionary biology, through which we attempt to read the Burgess Shale.

Outside paleontological circles, the Burgess Shale is known primarily for its starring role in the popular book *Wonderful Life*, by the late Stephen Jay Gould.[17] In *Wonderful Life*, Gould emphasized the huge diversity of life-forms represented in the Burgess beds. The fauna are not merely diverse (as measured by the number of species present) but, Gould argued, *disparate*, in the sense of having a wider variety of basic body plans than the animal kingdom has today. The Burgess fossils include representatives of not only all modern animal phyla save one (a phylum being the highest taxonomic category within a kingdom), but also several other phylum-level lineages that no longer exist. Even our own phylum, Chordata (of which vertebrates are the primary subdivision), is represented, by a diminutive, eel-like creature called *Pikaia*. Gould used the Burgess Shale to illustrate what he believed to be larger evolutionary truths: (1) that survival is as much a matter of luck as of fitness and (2) that the modern spectrum of flora and fauna represents only a rather narrow slice of a much wider range of possible, and equally functional, designs. Gould challenged the standard depiction of the tree of life—gradually branching and ever widening—and suggested instead that over time life is more like a bush, sending out many branches all at once. After each period of growth, most branches get pruned by extinction, but a few lucky ones go on to define later lineages.

Gould's interpretations of the Burgess Shale have been attacked rabidly by some members of the paleontological community, including Simon Conway Morris, whose graduate thesis work on the Burgess Shale was extolled in *Wonderful Life*. Conway Morris and others now maintain that the Burgess animals were not quite as bizarre as their own earlier work had suggested.[18] According to Conway Morris, several species that were once icons of weirdness, considered to lie entirely outside known taxonomic categories, actually do have family ties to known lineages. *Wiwaxia*, for example, a most peculiar creature resembling either the helmet of Hermes or a winged artichoke, is now interpreted as a primitive brachiopod, a type of two-shelled marine organism that still lives in modern seas. Otherworldly *Anomalocaris*, now renamed *Laggania cambria*, has been reinterpreted as just another arthropod, not all that different from crabs and lobsters. (Most of the fossil species in the Burgess Shale belong to the phylum Arthropoda, which includes modern insects and remains one of the most successful lineages of all time.) In a complete reversal of his own earlier conclusions, Conway Morris is now reluctant to assign any of the Burgess creatures to "new" (actually old) phyla.

Other evolutionary biologists, most notably Richard Dawkins, have challenged the significance and even the reality of the so-called Cambrian explosion, the term geologists use to describe the burst of biological creativity recorded in the Burgess Shale.[19] In his 1990 review of *Wonderful Life*, Dawkins simply brushed aside the importance of the large number of phyla in the Burgess Shale, pointing out quite correctly that "every new phylum has to start as a new species."[20] That is, the first bifurcations in a branching system will necessarily form the limbs from which twigs later grow. As a geologist, I feel compelled to point out that Dawkins misses the key point—that the bifurcations in the tree of life started late and suddenly, after the stout trunk had already been around for more than 3 billion years. Whether one is counting species or phyla, something big happened in the middle Cambrian.

The fierceness of the debate surrounding the Burgess Shale and the Cambrian explosion suggests that there are larger philosophical and political agendas behind the science. Some of the vehemence in the objections to Gould's ideas comes from the unflagging vigilance evolutionary biologists must maintain against creeping creationism. Gould was certainly no creationist—indeed the central purpose of his voluminous and sometimes bombastic popular writings was to expose as many readers as he could to the logic of evolutionary thinking.

Gould did challenge some tenets of orthodox Darwinism, however. In particular, Gould's paleontological work on trilobites led him to conclude that evolution did not in general proceed at a steady, stately rate but sometimes leaped ahead and then rested again, in a pattern he and colleague Niles Eldridge dubbed **punctuated equilibrium**.[21] The provocative subtitle of the 1977 paper in which the idea was first published was "The Tempo and Mode of Evolution Reconsidered." Staunch Darwinian watchdogs could hear the creationists rubbing their hands in glee. Punctuated equilibrium seemed to play directly to creationist "missing link" arguments—that the apparent absence of bridging organisms in the fossil record disproved evolution. From the perspective of evolutionary biologists, it was much safer to attribute the "sudden" appearances of species at different times in the geologic past to the highly imperfect nature of the fossil record.

But since the early 1980s, methodical paleontological research, together with much higher resolution isotopic age dating, has shown that the fossil record is not always so fatally flawed. Many alleged missing links have been found (e.g., dinosaur to bird, land mammal to whale), and it has also become clear that evolutionary rates are simply not constant. Comparatively rapid *radiations*—periods of branching and experimentation—have occurred many times in the geologic past, particularly after episodes of devastating environmental change and mass extinction (e.g., after the Permian–Triassic debacle and the Cretaceous–Tertiary meteorite impact). So the idea of punctuated equilibrium has been somewhat grudgingly accepted by the paleobiological community.

Gould's arguments about the rapidity of evolutionary progress recorded in the Burgess Shale, however, obviously struck a particular nerve with ultra-Darwinists like Dawkins. (Whether Darwin himself would feel comfortable in this group is open to question.) My own view is that this disagreement reflects a cultural divide between geologists and biologists. Geologists are always somewhere in geologic time, immersed in the idiosyncratic particulars of different periods in Earth's past. Biologists are of course aware of the deep evolutionary history of organisms, but see evolution mainly from the temporal plane of the present. The successes of molecular biology, moreover, have led to a seductive view of organisms as machines that should behave as predictably as clockwork. If organisms could live and evolve in front of an unchanging environmental backdrop, then perhaps evolutionary rates would be nearly constant. But on a planet prone to changes, not only slow and cyclical but also sudden and singular, why would we expect evolution to keep a steady beat? If it had, there would probably be no organisms around to tell, or argue about, the tale. From a geological perspective, the capacity to change—fast—when confronted with environmental challenges would seem the very best kind of evolutionary fitness. The rest of the time, it makes sense simply to hold one's own.

A second, and somewhat even more fundamental, point of contention between Gould and Conway Morris, Dawkins, and others is the extent to which modern organisms, especially animals, constitute the best possible designs or are nonoptimal artifacts of history, like the QWERTY keyboard.[22] The disagreement is deep and divisive because it bears on our interpretation of ourselves. In Gould's view, modern animals, as diverse and cleverly adapted as they (we) are to various environments, simply represent the best adjustments that could be made to the genetic legacies of the middle Cambrian evolutionary lottery winners. If the tape could be rewound, and if the "exotic" Burgess Shale animals had instead survived, then the path of evolution might have been utterly different.[23]

Conway Morris, in contrast, places great emphasis on the phenomenon called **convergent evolution**—the process by which the same features or body shapes have evolved independently on different branches of the tree of life.[24] Classic examples include the remarkable similarities between the shapes of modern dolphins and Jurassic swimming reptiles called ichthyosaurs, or the parallelism in the types of marsupials in Australia (long isolated tectonically) and the placental mammals in the rest of the world. Such examples suggest that, for some environments, there is a single optimal design and that natural selection will patiently sculpt organisms into that form. The logical outcome of this line of reasoning, advocated by Conway Morris, is that evolution of humans or humanlike organisms was in some sense inevitable since our attributes represent the "best of all possible" forms.

This Panglossian and self-aggrandizing argument is rather out of character for the otherwise clear-eyed Darwinists. To me, it is a bit like someone's declaring the fork the best possible eating utensil without being aware that half the world does quite well with chopsticks. We have no other world with which to compare evolutionary outcomes; how do we know that there are no other equally"good"possibilities? But Gould's view of evolution as a little more than a random walk seems excessively self-deprecating. Surely the course of life on Earth, like the course of any individual life, has been shaped by both the systematic (relatively constant natural selection pressures or cultural norms) and the stochastic (blind luck or bad karma).

AN ARTHROPOD-EAT-ARTHROPOD WORLD

Let us now return to the seafloor of the middle Cambrian, where there is less existential angst but plenty of bloodshed. One aspect of the Burgess Shale about which there is consensus is that predation was well established by that Cambrian day when the Burgess creatures happened to become entombed within a submarine mudslide. Well-developed digestive systems and hard exoskeletons show that these organisms had learned both to eat and to avoid being eaten. Many of the organisms are so well preserved that we can identify their stomach contents to the species level. *Anomalocaris* (more recently called *Laggania cambria*), at up to 1 1/2 feet long, was the top carnivore, with a choice of more phyla than there are selections at a modern sushi bar.

We see no fossil beds in which Ediacaran and Burgess Shale types of creatures coexist and therefore no direct record of what happened to the peaceable kingdom of Ediacara. But the fact that the very last strata with Ediacaran biota date from 543 million years ago, and that some Burgess-like organisms appear in the Chengjiang fossil beds in China, formed about 520 million years ago (somewhat earlier than the Burgess Shale), is suggestive.[25] (The Chengjiang strata, like the Burgess Shale, are Lagerstätten beds in which soft body parts were preserved as a result of rapid burial that spared the bodies of the animals from the ravages of oxygen.) The Chengjiang beds include a chordate even more primitive than *Pikaia* (*Cathymyrus diadexus*, or the "Chinese eel of good fortune") and also record rampant carnivorous feasting. Either the Ediacaran organisms evolved into some of these gastronomes via as yet undiscovered evolutionary pathways, or they were gobbled up when a new line of animals with this dangerous new habit appeared in their waters. A few Ediacarans, such as the doily-like *Kimberella*, have been tentatively linked to later lineages.[26] In any case, the meat-eaters must have had progenitors. Perhaps they are among the cryptic "small shelly fauna" of the lowermost Cambrian (these are the tiny shells that greatly troubled Darwin since older beds appeared to be entirely barren of fossils). Although the phylogenetic affinities of the "small shelly fauna" are still being resolved, the simple presence of their hard external armor indicates that they lived in a world where self-defense was necessary.[27]

Predation is, of course, a brilliant strategy for survival; rather than having to gather diffuse nourishment directly from seawater (the ultimate slow food), one can quickly ingest large bites of high-quality protein that has been conveniently preprocessed by other organisms. Predation is neither good nor bad in any absolute sense, but once introduced, it utterly changes the rules of an ecosystem. Carnivores can eat (but otherwise ignore) vegetarians, but vegetarians can neither eat nor ignore carnivores. It is the same kind of asymmetry that arises when monotheistic colonial powers meet polytheistic local religions: Polytheism can tolerate monotheism, but not vice versa, and so monotheism inevitably becomes the ruling system. This phenomenon is also akin to the *tragedy of the commons*, a phrase first used in an 1833 pamphlet by amateur mathematician W. F. Lloyd.[28] The idea is that as long as no one allows too many animals to graze on com-

munally held pastureland, a group of farmers can sustain themselves, but as soon as one person puts an extra animal on the
land, the others see no reward for restraint. In Lloyd's words, "The
essence of dramatic tragedy is not unhappiness. It resides in the
solemnity of the remorseless working of things."[29]

The early days of predation must have been remorseless. But
this new world order also led to unprecedented innovations: not
only protective shells and exoskeletons, but legs, backbones, eyes,
and champion swimming.[30] There was no turning back; life on
Earth was irrevocably changed. Remarkably, the evolutionary innovations sparked by the early Proterozoic oxygen revolution and
those in the Cambrian explosion have antipodal origins: in the
first case, intimate collaboration and symbiosis; in the second,
ruthless competition.

The Many Legs of the Arms Race

In some ways, the rest of the story of life on Earth is simply
about tinkering with designs that had been established in the
Cambrian. Climates changed, continents drifted, meteorites
struck, and at the end of the Permian, the whole biosphere
nearly succumbed to the oceanic equivalent of blood poisoning
(Chapter 4), but after each upheaval, at least one of the old lineages gained ascendancy again. In the Mesozoic, it was the reptiles; for most of the Cenozoic, it has been the mammals. The
birds even had their brief day in the sun during the earliest period of the Cenozoic era, when the nightmarish *Diatryma*—eight
feet tall, with gigantic hooked beaks—reigned as the top predators on land.

Each of these chapters recounts some variation on the theme of
an arms race. Predator and prey are locked together in an intimate dance with increasingly complex choreography, matching
each other step for step, neither gaining any actual ground. In

Devonian time, fearsome armored fish invented jaws, and their prey grew better fins for evasive escapes. In the Jurassic, as carnivorous dinosaurs got bigger and toothier, the plant-eaters grew sharp spikes and clublike tails. Evolutionary biologists call this the Red Queen effect, a reference to a character in Lewis Carroll's *Through the Looking Glass.* The Red Queen tells Alice that "in this place, it takes all the running you can do to keep in the same place." Neither predator nor prey becomes fitter vis-à-vis its contemporaries, even though both "improve" over time.

If the Red Queen effect seems depressingly Sisyphean, it is also a stabilizing phenomenon, whose effect is to maintain the trophic structure of ecosystems. But the principle applies only to predators and prey that have evolved in the same relatively stable matrix, and it occurs only as long as predator and prey evolve at about the same rates. If any of these conditions does not hold true, one side or the other may go from running in place to running amok. This is a lesson that has implications not only for wildlife management but also for agriculture and medicine. In trying to vanquish our insect foes, we have inadvertently created superior versions of them (while also poisoning our own food, water, and soils). Similarly, the continued use of antibiotics at current rates will only lead to increasingly virulent strains of bacteria by imposing far stronger selective pressures on disease-causing microorganisms than nature would. Because both kinds of "bugs" can reproduce and therefore evolve many times faster than humans, we will never keep up with insects and microorganisms. That is, we have increased the pace of the Red Queen treadmill and are now desperately trying to maintain our footing.

The modern problem of invasive species illustrates what happens in the absence of the Red Queen effect. When an exotic plant or animal arrives from distant shores without its coevolved rivals and enemies, its treadmill becomes a moving walkway that allows it to travel across the landscape much faster than the native organisms, which are plodding along in place. Importing the corresponding predators often wreaks even more environmental havoc. Introducing genetically modified organisms, which are created entirely outside an evolutionary context, could have even more devastating consequences. In a natural system that has evolved over time, each creature has its nemesis; predator and prey, parasite and host are well-matched rivals. Ecosystems are tightly woven nets of coevolved species that cannot readily be dismantled, reassembled, or fabricated without serious consequences for the integrity of the whole.

Arms races in the geologic record always end, but never with victors. Instead, an external referee—a meteorite, an ice age, a methane belch—abruptly changes the criteria for fitness, and all the elaborate armaments and defenses so assiduously stockpiled become as useless as a credit card in the wilderness. Then it is a matter of finding new uses for the specialized machinery developed under the old regime.

COMMUNES AND JUNKYARDS

For every arms race chronicled in the fossil record, there are counterexamples of symbiotic coevolution. In the Silurian period, about 425 million years ago, for example, the first modern reefs emerge—coral cities hosting dense and diverse populations of marine organisms. These reefs provided new kinds of domiciles and lifestyles not possible on the open seafloor. The multitiered reef structure provided organisms with access to sunlight, more sources of food, and protection from predators. In the Great Lakes region, some of these ancient reef structures can still be seen in situ (a spectacular example is in the parking lot of old County Stadium in Milwaukee).

The reefs continued to influence the environment for millions of years after the seas retreated. As sea level fell in the later Silurian, the reefs acted as barriers that isolated the central part of the shallow basin (what is now lower Michigan and Lake Erie). The waters became increasingly briny, and immense amounts of salt precipitated out over time. These salt beds, locally more than a thousand feet thick, have long been mined in Michigan and northern Ohio. Interestingly, there is a correlation between exposures of salt beds and finds of ice-age mastodon bones. Apparently, the beasts sought out these outcrops as natural salt licks more than 10,000 years ago. It is a lovely, almost literary image: tundra animals sustained by the remains of an ancient tropical marine civilization. Even today, the Silurian reefs shape the landscape. The dolomite beds in which the reefs lie are much more resistant to erosion than are the underlying Ordovician shales, and the result is an abrupt step in the landscape, the Niagaran escarpment. The escarpment wraps around the three inner Great Lakes, from Door County in Wisconsin through the Upper Peninsula of Michigan, into southern Ontario and back down to New York, where it defines the position of Niagara Falls.

At the same time that the great underwater cities of the Silurian were being built, other organisms were moving out to the unpopulated suburbs—onto the land. Although cyanobacteria and other microbes had been living in terrestrial freshwater environments as early as Proterozoic time, macroscopic land ecosystems as we know them today did not emerge until the Silurian. Plants—first mosses and then various types of vascular species like horsetail and ferns—homesteaded in this open frontier. Leaving the water meant having to develop new strategies not only for staying hydrated but also for body support, temperature regulation, food gathering, and reproduction. These early nonmarine plants literally tamed the land for the first terrestrial animals—mainly insects and early spiders—which benefited from the shelter, food, and humidity and temperature control the plants provided.[31] By early Carboniferous time, 350 million years ago, there were forest ecosystems as rich in diversity as their marine precursors, with thousands of species enjoying the advantages of communal living.

Another instance of a glorious partnership in the geologic past is the pas de deux between insects and flowering plants beginning in the Cretaceous period, about 130 million years ago. The origin of flowering plants, or angiosperms, from the older, reproductively simpler gymnosperms remains somewhat unclear (Darwin called it "that abominable mystery"). Whatever their origins, once the angiosperms arranged with insects to provide food in exchange for pollination, the results were spectacular. Cretaceous clay beds in New Jersey and elsewhere record the simultaneous and explosive diversification of insects and angiosperms, a testament to the power of collaboration as a creative force in evolution. Today, angiosperms constitute more than 99 percent of all plant species (there are an estimated quarter million types of angiosperms and only about a thousand gymnosperms). Even more astonishingly, the number of insect species is larger than the total number of species in all other branches of the animal kingdom *combined*.[32]

Some innovations in the evolutionary past seem difficult to explain without resorting to **teleological** (directed, or goal-oriented) explanations. The origin of the wings and feathers in birds, for example, has been one of the classic conundrums in evolutionary theory. The puzzle is that half a wing is no good. Yet how could the wing—or any other complex feature—emerge fully fledged (literally, in this case)? The answer must lie in *polyfunctionality,* the potential for a structure to serve more than one function, like a scrap scavenged from a junkyard and put to a new use. The recent discovery of a four-winged, feathered dinosaur-bird hybrid in Liaoning, China, supports the idea that flight began with tree-to-tree gliding by four-legged animals (rather like the gliding of modern flying squirrels).[33] (Interestingly, a zoologist hypothesized in 1915 that such a "tetrapteryx" might have been a transitional species between birds and dinosaurs, but his theory was pecked to death.)[34] Feathers probably evolved from reptilian scales and appear to have been present in some dinosaur families long before they ever became useful in gliding or flight—perhaps serving originally as heat regulators.[35] Clearly, versatile parts that can be retooled to suit another purpose confer a great advantage on their owners.

Paradoxically, the capacity of organisms for great innovation may come from extreme conservation—keeping old spare parts and remembering techniques employed in the past. Evolutionary molecular biology has become an important complement to paleontology in understanding the interplay between the quest for stability and the imperative to change. Increasingly detailed genomic sequencing of organisms, including humans, has revealed vast stretches of junk DNA, that is, repetitious and redundant segments of the genetic code that have no clear use. Such units of DNA, known as *introns,* take up more than 95 percent of the space on human chromosomes. Some of these "unneeded" segments have been found to contain marker sequences that can be used to predict an individual's susceptibility to diseases. (As a result, there is ferocious rush, reminiscent of the dawn of predation in the Cambrian, to patent introns.[36] Such a practice seems as absurd and unethical as patenting a mineral or a species of trilobite.) More recently, a group from Harvard Medical School has challenged the idea that our cells would go to the bother of lugging around entirely useless baggage.[37] The researchers found that a junk gene on the yeast genome can act to switch off the function of an adjacent gene.[38] If this is true of other introns, then junk

DNA may actually be a powerful mechanism by which organisms maintain genomic flexibility. Perhaps our genes remember the lessons of the past and hold them in reserve for an eternally uncertain future.

SOMETHING OLD, SOMETHING NEW, EVERYTHING BORROWED

Our evolutionary past is full of paradoxes. The fossil record shows that collaboration is as potent as competition as a source of evolutionary innovation, and yet both concerts and contests can also foster stability. Endosymbiotic mergers between microorganisms invented the first complex cells. Sex made genetic experimentation more efficient. Predation inspired the body plans shared by most of the animals in the modern world. Living communally in reefs or in land ecosystems provided new opportunities for upward mobility but also carried its dangers.

Conservation and innovation are curiously intertwined in evolution. Reproduction strives for high-fidelity copying of well-tested genes, but errors and anomalies make improvement possible. Success seems to require walking a knife edge between caution and risk taking. Stick with a trusted formula—but don't become complacent! Understand the past, but be ready for change. Keep spare parts at hand.

The tension between the old and new reflects the dichotomous character of geologic time: the cyclical and recurrent versus the linear and singular. Some rules that govern organisms never vary (e.g., gravity), some cycle slowly over time (climate, usually), some change steadily and irreversibly (the biosphere itself), and some are brief but devastating decrees (meteorite impacts). But no organism has ever evolved outside the rules of a physical and evolutionary environment.

Human consciousness is arguably the first truly novel innovation to arise since Cambrian time, in the sense that the technologies our consciousness has spawned have freed us from the limits of our own body architecture. Perhaps even more important, consciousness has allowed us to transcend our single moment in geologic time and glimpse the great tapestry of Earth's history. But we are only beginning to understand the richness of Earth's history and the imprint it has left deep in our genes. For all of its potency, consciousness can become a pathological condition if it gives us delusions that we have somehow been exempted from the rules that have always governed the biosphere. The belief that we can engineer what evolution has done in 4 billion years—and expect the results to be predictable and controllable—is a sign of our youth and ignorance. Naive tinkering with such ancient systems is foolish, arrogant, and dangerous.

6

STRENGTH AND WEAKNESS

*He who conquers others has force; he who conquers himself has
strength.*

— LAO TZU, *TAO TEH CHING*

IS NATURE BENEVOLENT, malevolent, or neutral? Ultimately comprehensible or infinitely complex? Predictable or
chaotic? Beautiful or repugnant? Robust or fragile? The answers
have varied over time. A retrospective of how scientific perceptions of the Earth have changed over the past three centuries reveals a strong correlation between Western political and social
views and contemporary scientific "truths." This connection
should make us suspect that our understanding of the planet at
any historical moment is at best incomplete and at worst hopelessly wrapped up with our own self-image.

The limitless, chaotic, and threatening Earth of the early eighteenth century gradually became the measurable, mechanical,
tamable Earth of the nineteenth. At a time when society was
shackled by the struggle for survival, science was confident that

---

This chapter appeared in a slightly different form in *The Earth Around Us: Maintaining a
Livable Planet* edited by Jill Schneiderman (Westview Press 2003).

Earth could be tamed and compelled to yield up unprecedented bounty. This seductive vision continued to prevail through much of the twentieth century, but then the Earth began to become fragile and finite, requiring husbanding and management. Now a new scientific view of Earth is emerging, and ironically, it returns in part to the prescientific imagery of the 1700s. This Earth is again immeasurable (infinite in its complexity at many scales), unknowable (chaotic in the mathematical sense), and indomitable. And, in another irony, science and society have reversed their roles. The now-entrenched consumer culture acts like it owns the planet, while science is beginning to understand that our occupancy is recent and probably temporary. The planet, meanwhile, seems to be taking its own measures to remediate the damage caused by the past 300 years of reckless tenancy. Has science got it right this time? Why did it take so long to rediscover things we knew long ago? Have we learned anything at all? And what do we do now?

EARTH BEFORE GEOLOGY

By the late seventeenth century, the scientific revolution was well under way. Newton had reined in the heavens by tethering them to the same fundamental laws that operate on Earth. But the Earth itself remained untamed. Geology had not yet emerged as a distinct discipline with established premises and practices, and most scientific treatises on the formation of the Earth and its landscapes were baroque variations of the biblical account. The cataclysms that had shaped the Earth were depicted in imaginative detail in works like the illustrated *Sacred Physics,* by Johann Scheuchzer, a synthesis of naive paleontological observations, quasi-scientific reasoning, and biblical literalism.[1] To Scheuchzer and others, Earth's present was disjunct from its past. Different rules had applied in primordial times, and in any case, the rules were not necessarily meant to be comprehensible by the human intellect. The forces of nature were capricious, controlled by the hand of an easily angered, if more recently mellowed, God.

Wilderness was considered brutal and monstrous. Mountains were pathological—"carbuncles on the face of the Earth"—a diagnosis very different from the majesty attributed to them today.[2] According to *Telluris Theoria Sacra* ("The sacred theory of the Earth"), a series of volumes written in the 1680s by influential theologian and natural philosopher Thomas Burnet, mountain ranges and canyons were the scars left on a once perfectly smooth "mundane egg," which cracked and released the Noachian deluge.[3] Burnet's geological assertions appear ludicrous to modern readers and were treated with particular scorn by Charles Lyell as early as 1830 in his review of theories of the Earth.[4] But as Stephen Jay Gould argues, Burnet's approach was scientific, or at least rationalistic, in that he was attempting to integrate sacred and secular knowledge into a single, internally consistent narrative. To Burnet and his contemporaries, the roughness of Earth's landscape was an expression of, and punishment for, human iniquitousness.

Within this inclement physical and spiritual landscape, humans were left to fend for themselves. In preindustrial times, the use of animal, vegetable, and mineral commodities (not yet called resources) was a matter of local contingency rather than systematic exploration and distribution. Without any understanding of the evolution of the landscape or the genesis of ore deposits, the discovery and extraction of usable Earth materials depended on serendipity. The cultivation of land was one of the few activities through which humans could exert direct control over the type and quantity of goods produced, and so agriculture-based social structures became the templates for economic and political systems in Western Europe.

John Locke's *Second Treatise of Government* (1689), a seminal document in both British and American political history, articulated the late seventeenth-century, agri-biblical vision of "natural" rights to land and other property. In Locke's view, the Earth was a divine gift to man, but no Eden. He depicted nature as a grudging provider, asserting repeatedly that land has no intrinsic value until human toil is invested in it:

> 'Tis labor that puts the greatest part of the value upon the land, without which it would scarcely be worth anything. . . . Nature and Earth furnished only the almost worthless materials, as in themselves.[5]

Moreover, Locke argued, it is God's intent that land be developed:

> God gave the world to men in common; but since he gave it them for their benefit and the greatest conveniences of life they were capable to draw from it, it cannot be supposed he meant it always should remain common and uncultivated.[6]

And when one man cultivated a plot of land, the society as a whole benefited:

> [H]e who appropriates land to himself does not lessen but increases the common stock of mankind. . . . For the provisions serving to the support of human life produced by one acre of enclosed and cultivated land, are . . . ten times more than those which are yielded by an acre of an equal richness, laying waste. . . . And therefore he that encloses land, and has greater plenty of the conveniences of life from ten acres than he could have from a hundred left to Nature, may truly be said to give ninety acres to mankind.[7]

In Locke's view, Earth's gifts could be won only through toil and sweat, but these riches were essentially infinite. Human labor was the sole limitation on resource availability, and humans had a moral and social mandate to cultivate, subdue, and domesticate. The concepts of extinction and exhaustion were not yet in circulation.

NAMING NAMES AND MAKING MAPS

Carolus Linnaeus, the great eighteenth-century Swedish biologist, set the agenda for other branches of natural history by organizing the biological world into an orderly hierarchy in which every organism had a name and a place in the pedigree of life. The quasi-divine act of naming things is an empowering and satisfying task, and taxonomy (together with a fair amount of taxidermy) became a preoccupation of the emerging natural sciences in the eighteenth and nineteenth centuries. The spirit of the scientific times was embodied in the natural history museums of the Victorian era. The buildings were bursting with stuffed birds, skeletons, fossils, crystals, and other natural wonders, named and tamed, entombed within glass cases.

In the geological realm, there was so much to classify—rocks, minerals, fossils, landforms, ore deposits, folds, faults—that the job continued well into the twentieth century. In the absence of unifying genetic models for the formation of most of these features, classification schemes brought a comforting sense of finitude and fixity to nature's variability. Some geologic entities, like minerals, fell easily into well-defined categories, and the great nineteenth-century encyclopedic treatises about them—for example, the mammoth *Manual of Mineralogy*, first published in 1869 by Yale professor James Dana—are still in use, the most complete compendia of their kinds ever produced.[8] But other geologic phenomena resisted ready classification, and the best minds of the time struggled to identify idealized Platonic categories that would impose structure on unruly reality.

Developing a universal nomenclature for ore bodies was a particular priority because of their economic importance, but this proved (and to some extent remains) notoriously difficult. The classification schemes were invariably entangled with idiosyncratic theories of ore genesis based on mineral occurrences in a particular region.

In the late eighteenth century, mining academies proliferated across Europe, with particularly vigorous programs in Germany, Sweden, and France. Many of these programs were headed by a single visionary surrounded by disciples who helped to propagate their master's system.[9] In Freiburg, the "neptunist" school, led by the imposing Abraham Werner (1750–1817), advanced the notion that all rocks and ore deposits were precipitated from seawater. In Uppsala, Johan Wallerius (1709–1785), an adherent of the ancient alchemical belief in the transmutation of metals, advanced the very modern view that chemistry, not external properties like color, was the key to understanding ore minerals. In Paris the apothecary-turned-mineralogist Guillaume François Rouelle (1703–1770) and his (more famous) student, the chemist Antoine Lavoisier (1743–1794), also developed early theories about the nature and distribution of rocks and ores.

Few of these theoretical schemes influenced actual mining practices, however; miners' experience and intuition generally proved more reliable. But the very existence of the mining academies signaled an important new philosophy: that the Earth and its mineral resources could be analyzed and ultimately understood.

James Hutton's doctrine of uniformitarianism (Chapter 2), based on his observations of rocks in Edinburgh and the Scottish borders, further helped "rationalize" the Earth. His interpretation of the unconformity at Siccar Point and "discovery" of deep time showed that Earth had been governed by the same physical laws in the past as in the present. (Hutton's observations also documented the existence of igneous rocks, a blow to Werner and the neptunists.) Hutton's work seemed to confirm that the Earth's behavior was logical, dependable (if not monotonous), and comprehensible:

> Philosophers, observing an apparent disorder and confusion in the solid parts of this globe, have been led to conclude, that there formerly existed a more regular and uniform state, in the constitution of this earth; that there had happened some destructive change; and that the original structure of the earth had been broken and disturbed by some violent operation, whether natural, or from a supernatural cause. Now, all these appearances, from which conclusions of this kind have been formed, find the most perfect explanation in the theory which we have been endeavouring to establish. . . . [T]here is no occasion for having recourse to any unnatural supposition of evil, to any destructive accident in nature, or to the agency of any preternatural cause, in explaining that which actually exists.
>
> We have the satisfaction to find that in nature there is wisdom, system, and consistency. . . . The result of our present enquiry is, that we find no vestige of a beginning,—no prospect of an end.[10]

This epoch of scientific elucidation coincided with (and was used to justify) European colonization and settlement of the Americas, Africa, and the southern Pacific. Expeditions were sent forth to document the flora, fauna, and mineral riches of the frontiers. The richly detailed, carefully illustrated notebooks of the 1803–1806 North American expedition by Meriwether Lewis and William Clark represent the best of these officially commissioned reports. In the United States, federal and state geological surveys were established for assessing and mapping the resources of the nation. These agencies were charged with census-taking—counting and accounting, making the infinite and indefinite finite.

In the Seventh Annual Report of the Geological Survey to the U.S. Congress in 1888, Director John Wesley Powell, who himself had led a great geologic expedition in the American West, wrote of the strategic importance of a new program to create accurate topographic and geologic maps of the nation.[11] Powell, the first to chart the lower Colorado River and the Grand Canyon, appreciated the power of maps. Like classification schemes, maps confer to the user a kind of ownership over their subjects. Maps miniaturize wilderness so that it can be held in the hand and seen in the mind's eye.

Maps and surveys were critical in the implementation of the U.S. Homestead Act of 1862 and the General Mining Law of 1872, both based on Locke's principle that anyone who worked a parcel of land (and could define its boundaries) was the rightful owner of it. The Homestead Act survived well into the twentieth century (it was repealed in 1977!), and the General Mining Law remains in effect today. An outrageous legal anachronism, the General Mining Law still allows anyone to search for and extract minerals on publicly held lands for a fee of less than five dollars per acre, with little accountability for any resulting environmental damage. These federal policies and the philosophies behind them were also responsible for some of history's most egregious instances of systemic social injustice: the repeated reneging by the U.S. government on treaties with Native American tribes.[12] These doctrines were intellectual extensions of Locke's value-added principle of property rights: that those who could fathom nature, name it, and map it have the right to exploit it. Aboriginal systems of naming and knowing, however, were not recognized as legitimate.

A MECHANICAL EARTH

By the early nineteenth century, mechanization had begun to change the balance of power between people and the planet. As machines proliferated, there seemed no limit to their potential to magnify human strength. The imagery of the machine, the mindless automaton, was both thrilling and frightening, a powerful subconscious metaphor for things only partly understood. It was natural, therefore, that once its parts were identified and it operations began to be comprehended, Earth too would be viewed as a great machine.[13] (The metaphor of mechanism was well enough established by the early 1800s to spawn the cultural counterreaction of Romanticism, a celebration of all that was wild and amechanistic in nature.) The cyclicity implicit in James Hutton's doctrine of uniformitarianism resonated with the cogs and flywheels of the age of steam (Hutton was in fact a friend of James Watt, inventor of the steam engine). Although Hutton advocated the cyborgic phrase *living machine* to describe his vision of a self-renewing, self-repairing Earth, the embarrassing organic adjective was dropped by his followers, including Charles Lyell. In his *Principles of Geology* (1830), Lyell characterized the workings of the Earth as a proud factory owner might describe his manufacturing machinery:

> It is to the efficacy of this ceaseless discharge of heat and of solid as well as gaseous matter, that we probably owe the general tranquillity of our globe; for if it were not that some kind of equilibrium is established between fresh accessions of heat and its discharge, we might expect perpetual convulsions. . . . But the circulation of heat from the interior to the surface, is probably regulated like that of water from the continents to the sea, in such a manner that it is

only when some obstruction occurs that the usual repose of nature is broken.[14]

Lyell's Earth machine was benign and efficient, humming along in a steady state except on rare occasions, when something jammed. Even then, normal operations would soon be restored. Such mechanical imagery promised a new level of control over nature. If one could only understand the mechanism, the outcome would be determinable. This view was most confidently advanced by mathematician Pierre-Simon Laplace in his 1814 *Essai Philosophique*. He asserted that if all forces and bodies in nature could be accounted for, "nothing will be uncertain, and the future, as well as the past, would be present to the eyes."[15]

Later in the nineteenth century, knowledge of the future began to look somewhat less attractive. With the rise of thermodynamics, it became clear that Earth's life span was limited, that eventually its great engine would run down and grind to a halt. Lord Kelvin's infamous (mis)calculation of the age of the Earth (Chapter 1) also implied a finite future, as the Earth evolved inexorably toward thermal equilibrium.[16]

The mechanical model for the Earth has been undeniably successful in many ways, allowing us to see recurrent patterns within the idiosyncratic particulars of the Earth's history. In the middle of the twentieth century, the Earth machine became somewhat more complex, with a new set of moving parts in the form of drifting continents. In most histories of geology, the plate tectonic revolution is depicted as an epiphany, a scientific coming of age, the paradigmatic paradigm shift. It would be absurd to deny the intellectual, and practical, importance of plate tectonics. Economically, for example, tectonic theory provided, at long last, the unifying framework for mineral resources that Werner and others had sought. Yet in terms of underlying metaphor, plate tectonics was only an elaboration on a nineteenth-century theme.

The incredible shrinking Earth

At the beginning of the nineteenth century, Thomas Robert Malthus, in his *Essay of the Principle of Population* (1798), had been a solitary voice warning of what today is called unsustainable resource use. Malthus's treatise was widely read, and almost universally criticized, throughout the rest of the century for its "inhumane pessimism" and "abominable tenet."[17] Friedrich Engels called it a "vile and infamous doctrine . . . repulsive blasphemy against Man and Nature."[18] Charles Darwin was among the few who were willing to consider the implications of Malthus's ideas. He realized that Malthus's central concept—that superfecundity leads to competition for resources—could provide the motive force for evolution by natural selection.[19] Although Darwin emphasized the effect of scarcity on the interactions among organisms, neither he nor most of his contemporaries were particularly concerned with the ultimate finitude and potential depletion of resources. In 1883, for example, Darwin's vocal disciple T. H. Huxley asserted his belief that the North Atlantic cod fishery was effectively infinite—that "natural checks" would always allow the resource to replenish itself "long before anything like permanent exhaustion has occurred."[20] Malthus's message was more than a century ahead of its time.

But as the Earth machine whirred into the twentieth century, it began to show signs of frailty that would have disconcerted Lyell. In the United States, the Dust Bowl of the 1930s was an alarming awakening to how quickly a critical resource could be exhausted. Farmland, which had so recently been robust and abundant, was suddenly sickly and scarce. Soil and water conservation became federal priorities. Resources now became entities to manage. The cherished illusion of limitless resources could no longer be maintained. Aldo Leopold called for farmers, voters, and policy makers to change metaphors and see the land as not a machine but an organism, a living entity with which we have an ancient relationship that carries ethical obligations:

> The land ethic enlarges the boundaries of the community to include soil, waters, plants, animals, or collectively: the land. . . . [It] presupposes the existence of some mental image of land as a biotic mechanism . . . [and] changes the role of *Homo Sapiens* from conqueror of the land-community to plain member and citizen of it. It implies respect for his fellow-members, and also respect for the community as such."[21]

But Leopold, like Malthus, was understood by few in his time. The exigencies of World War II and the hysterical consumerism of the 1950s made it possible to embrace the illusion of infinitude again. The publication of Rachel Carson's *Silent Spring* in 1962 is usually cited as the event marking the public's reawakening to the limits of nature's resilience. Through the rest of the 1960s and 1970s, the rhetoric of the environmental movement became progressively more pessimistic.

The titles of introductory geology textbooks from the 1970s reveal an almost claustrophobic preoccupation with Earth as a closed system (another nineteenth-century thermodynamic concept) and the adversarial relationship between humans and the Earth: *Man's Finite Earth; Earthbound; Geology: The Paradox of Earth and Man*. Popular titles were apocalyptic: *The Population Bomb; The Limits to Growth; Future Shock*.[22] The once bountiful and commodious Earth was suddenly too small. A popular geology textbook of the time began: "This book deals with the environmental problems caused by runaway population growth on a planet that has unalterable dimensions but fixed or shrinking resources."[23]

Apollo mission images of Earth as a blue ball floating in space underscored the feeling of finitude: "It's so incredibly impressive when you look back at our planet from out there in space and you realize so forcibly that it's a closed system—that we don't have any unlimited resources, that there's only so much air and so much water," said Edgar Dean Mitchell, an *Apollo 14* astronaut.[24]

Behind these words is the ancient fear that we will one day wake up to find the cupboards bare. An old and frail Mother Earth will no longer be able to provide for her children.

EARTH UNBOUND

But the apocalypse has been postponed. The global famines predicted in *The Population Bomb* for the 1970s and 1980s did not occur. *The Limits to Growth* forecast the depletion of copper and other strategic metals by the early 1990s, but reports of the dearth of these commodities appear to have been greatly exaggerated. In 1980, economist Julian Simon challenged ecologist Paul Ehrlich to a wager about how the prices of five metals would change in the coming decade. Ehrlich anticipated that increasing scarcity would drive prices higher over the subsequent decade. Simon predicted that prices would decrease, and he won the bet.[25] What happened to finitude?

The Limits to Growth, by the Dartmouth-based "Club of Rome," was groundbreaking in its use of systems modeling to grapple with complexly linked entities. The world model at the heart of *Limits,* with its profusion of circles and arrows, was then by far the most ambitious attempt to integrate natural, economic, and social variables into a single scheme. Like Burnet's *Sacred Theory of the Earth* nearly three centuries earlier, the world model was an earnest, if naive, effort to create a single, unified vision of Earth. The limitations of *Limits,* some of which were recognized within a year of its publication, arose largely from its faith in the nineteenth-century view that Earth is measurable and mechanical.[26] But the comforting concepts of measurement and mechanism were becoming unexpectedly problematic.

In 1967, Benoit Mandelbrot's concept of fractal geometry (Chapter 3) showed that something as simple as measuring a coastline can be tricky since the answer depends on the length of the ruler you use.[27] Coastlines seem to stretch as the yardstick gets shorter. How long will copper last? How many people can Earth support? It depends. If you set the price higher, copper will last a few more decades. If most earthlings became vegetarians, the planet could support another billion or two people. There's always more of what you're seeking if you look closer, pay more, accept substitutes, or bother to recycle. As Joel Cohen of Rockefeller University has argued, "Ecological limits appear not as ceilings but as trade-offs."[28] The feedback mechanisms of the global marketplace may give us a little more time. Resources remain finite, but as they grow scarcer, their value will rise—if, and only if, their market value reflects the true and comprehensive cost of their harvest or extraction. The apocalypse will not be a sudden jolt, but a gradual, inexorable squeeze. Unfortunately, people rarely respond to, or even notice, this kind of slow-motion crisis until it is far too late.

We may no longer have confidence about measuring coastlines, but appraising the planet might be possible. In an audacious 1997 paper in the scientific journal *Nature,* Robert Costanza and twelve other ecologists and economists from around the world suggested that monetary units might be the appropriate metric for an

Earth for which global market forces seem as strong and irresistible as the magnetic field. When everything was tallied, the estimated value of the "world's ecosystem services and natural capital" came out to $33 trillion—about twice the global gross international product. At first glance, such an analysis seems horribly crass. The point, however, is not that environmental decisions should be made strictly on economic grounds, but that the hidden environmental costs of economic activities—for example, fossil fuel combustion, suburban sprawl—should be acknowledged in the market. Costanza and his coauthors assert that their analyses have two types of practical application: (1) helping nations and international bodies modify "systems of accounting to better reflect the value of ecosystem services and natural capital" and (2) appraising governmental or commercial initiatives in which "ecosystem services lost must be weighed against the benefits of a specific project." They concede that the undertaking is at some level absurd:

> We acknowledge that there are many conceptual and empirical problems inherent in producing such an estimate. . . . The economies of the Earth would grind to a halt without the services of ecological life-support systems, so in one sense their total value to the economy is infinite.[29]

A growing number of environmentalists embrace the premise that the only viable solutions to global environmental problems will be those that harness the power of the global market.[30] In some sense, incorporating environmental entities into account sheets and annual reports reestablishes the rules of the pretechnological world, where calories were the currency and every member of the biosphere had to pay more—or evolve—when a commodity became scarce.

Aldo Leopold would have been skeptical; nearly sixty years ago, he argued that no system of economic incentives would succeed without "an ethic to supplement and guide economic relation to the land" and that the key to developing such an ethic was to "quit thinking about land-use as solely an economic problem." But economic policy can be implemented faster than new ethical systems, and perhaps by converting environmental metrics into currency units, behaviors, if not mores, will be changed. The contingent and imprecise nature of these monetary measurements must always be borne in mind, however.

At the same time that nature was beginning to resist measurement, the Earth machine was also showing signs of unpredictable behavior. In the tradition of mathematician Laplace, many modern scientists have continued to embrace the idea that knowing nature's rules—holding the user's manual—will eventually lead to omniscience. Biologist E. O. Wilson, for example, has recently described his "conviction, far deeper than a mere working proposition, that the world is orderly and can be explained by a small number of natural laws."[31] But even if the number of laws is small, the number of possible outcomes may be infinite, and order, if it exists, is probably ephemeral.[32] These are the lessons emerging from studies of Earth's climate, for example, which is better characterized by the mathematics of chaos and nonlinear systems than by Kelvin's equilibrium thermodynamics.[33] Even if we could build a complete model of the climate system and set it in motion, the output would not necessarily be meaningful, because of the number of variables that could change over the time of the simulation. The Laplacian vision of omniscience and certainty has given way to statistical probabilities and risk assessment. (Anyone who doubts the reality of global warming should check with insurance company statisticians.)

Similarly, our reading of the fossil record now tells us that biological evolution was never a stately march toward improved designs but a series fits and starts, buffeted by chance and punctuated by catastrophes.[34] Contrary to Lyell's view, nature apparently does not spend most of its time in repose. Rather, it is either running in place or responding to the latest environmental regime change. What happened to equilibrium?

In the study of some biological and geological systems, scientists now commonly invoke multiple equilibrium states separated by narrow thresholds, rather than a single, unique equilibrium state. In these models, even a relatively small perturbation to an apparently stable system can trigger systemwide reorganization, by touching off potent feedback mechanisms. The ocean's thermohaline circulation system, for example, can be switched on or off by relatively small changes in North Atlantic salinity. Similarly, climate modelers suspect that a critical temperature increase in arctic regions could cause peaty organic matter now stored in tundra soils to decompose rapidly. The resulting large volumes of greenhouse gases released would raise global temperatures further (a scenario eerily reminiscent of the ultra-greenhouse conditions that followed the Snowball Earth period and the Permian–Triassic extinction).[35] If it once seemed placid and domesticated, nature is now easily provoked and is biting back.

Some scientists have begun to question the validity of ascribing even this qualified kind of equilibrium to natural systems. Paul Rabinow summarizes this perspective: "Nature reflects the accumulation of countless accidents, not some hidden harmony. Things might have turned out quite differently. Ecosystems are ever changing, dissolving, transforming, recombining in new forms."[36]

Others argue that sentimental attachment to the idea of the balance of nature has undermined the credibility of environmental policy:

> The classical equilibrium paradigm in ecology . . . has failed not only because equilibrium conditions are rare in nature, but also because of our past inability to incorporate heterogeneity and scale multiplicity into our quantitative expressions for stability. The theories and models built around these equilibrium and stability principles have misrepresented the foundations of resource management, nature conservation, and environmental protection.[37]

If there is stability in nature, it is not static, but a far subtler dynamic, impermanent constancy.

The Earth machine has seemingly become too complex for the mechanics to understand. Is it time to acknowledge, with the Romantic poets, that the metaphor of mechanism fails to portray the full richness of nature? The geosciences appear ready to return to Hutton's half-organic depiction of the Earth as a living machine. There are surprising similarities between Hutton's late-eighteenth-century vision and James Lovelock's late-twentieth-century Gaia hypothesis, which places life at the center of the global climate system and geochemical cycles (Chapter 1).[38] Although Gaia has not yet been fully welcomed into the scientific fold, the goddess of the Earth has been ushered in through the backdoor in the less controversial guise of geophysiology.[39] In any case, Earth seems to have recovered its strength and spirit. The prognosis for the human race is less clear.

In many ways, scientific views of Earth have come full circle in 300 years. For centuries, humans feared the power and magnitude of a capricious and capacious Earth. Then we passed through a period of childlike self-aggrandizement in which we, like Saint-

Exupéry's Little Prince, were masters of a planet that was just the right size for us to explore and conquer. Rather suddenly we found ourselves surrounded by the wreckage of our careless expeditions, on an Earth that seemed to be shrinking. But if we just stop to look, we can glimpse infinity again, in every grain of sand and living cell on this old Earth—a planet that is at once benevolent and malevolent, comprehensible and complex, predictable and chaotic, robust and fragile.

Epilogue:

The Once and Future Earth

I would feel more optimistic about a bright future for man if he spent less time proving that he can outwit Nature and more time tasting her sweetness and respecting her seniority.

— E. B. White

W E   H U M A N S   C A N N O T   B E   B L A M E D   for wanting to believe that Earth was simple, predictable, and controllable, especially in a time when our own survival felt so uncertain. But now that we know the planet is none of these things, we are accountable to future generations for continuing to carry on as if it were. Although Earth science has come to recognize the wisdom in some older views of the planet, we do in fact know far more about it than we did in 1700. The most important finding, paradoxically, may be that some aspects of the Earth will always remain unknown. Our best maps will always be incomplete, our most precise age determinations will still have uncertainties, our search for origins will never lead to unambiguous starting points. And yet, these attempts to make sense of an immensely complicated world are not futile or meaningless as long as we bear in mind that they are only partial representations of something far vaster and deeper.

What do we know with any certainty? That we live on an old planet. That over geologic time, the face of the Earth has changed ceaselessly and yet on the whole has been a remarkably hospitable place. That Earth owes its unlikely equability to the balance of power between the commensurate forces of rock, water, and life. That stability is favored by efficient recycling and a well-mixed atmosphere and hydrosphere, by a diverse biosphere with many interacting tiers, by cooperation and competition, by innovation and conservation. We know that under certain rare circumstances, the Earth system can become wildly unstable and that although the planet has always recovered from such spells, recuperation can take millions of years. Extinctions, followed by explosive evolutionary innovations, have often followed these periods of global instability. We know too that the rates of human-induced changes in some of Earth's systems equal or exceed those associated with the most devastating geologic catastrophes. Current extinction rates, as measured by species per century, probably rival those of the greatest mass extinctions of the geologic past.[1] Present rates of change in atmosphere chemistry are likewise extreme, even by geologic standards.

And some types of anthropogenic changes simply have no precedent. A recent analysis of satellite imagery, for example, has shown that the total "constructed" area in the continental United States is now equal to the size of the state of Ohio.[2] Never has so much of the Earth's surface been covered by materials designed to be impervious (concrete, pavement, buildings).[3] These surfaces not only decrease the proportion of precipitation that soaks into the substrate to become groundwater, but also change the reflectivity, biological diversity, and carbon storage capacity of the land. Not all of these changes are necessarily bad, but they will interact in subtle and unpredictable ways with other environmental changes, both natural and human-induced. The next great evolutionary radiation may be of plants and animals adapted to the concrete world of the suburbs, a fast-growing environmental niche with plenty of opportunities for herbicide-resistant dandelions, supermosquitoes, and overgrown pigeons, raccoons, and rats with a taste for french fries. Meanwhile, our own offspring spend more time in malls and virtual computer worlds than in anything resembling a natural ecosystem.

The Frankenstein monster has come to symbolize the horrors that result when humans attempt to engineer life. But in Mary Shelley's *Frankenstein*, the true horror and tragedy arise from the monster's awareness that he has no kin, no memory, no sense of his own history. He shares no connection with any living being, and his inconsolable sadness and loneliness become destructive rage. Our interventions in the natural world have already altered the biosphere grotesquely. But our ignorance of nature threatens to cast *us* as the Frankenstein monster, bereft of a past, knowing nothing but our own appetites, emotions, and inventions.

I wonder sometimes what the geologic legacy of humankind will be, the composition of the stratum that will record our own late Holocene epoch. What will outlast us? Concrete, rusting steel, quite a bit of plastic, unidentifiable goop, nuclear waste, sediments with anomalous amounts of phosphorus, nitrogen, mercury, lead, and lots of isotopically light carbon from three centuries of burning fossil fuels that had taken half a billion years to accumulate. Rocks from all over the world jumbled together (thanks to the building stone industry and geologists and their collections). Evidence of a mass extinction.

What might we do to leave a more distinguished entry in the geologic record? For a start, we can spend more time conversing with existing rocks, whose gravitas may help us stop thinking in throwaway sound bites and anesthetizing euphemisms. Rocks will tell us that we can trust the Earth, that it is immensely old, durable, and infinitely wiser and more patient than humans. Rocks will remind us that arms races never have victors. Rocks may cause us to realize that mountainous garbage dumps, unopposed predation, unchecked consumption, and the flux of commodities from the poor to the rich are violations of ancient laws mandating recycling and redistribution. Rocks may even cause us to rediscover thoughtful discourse about complex environmental issues and to instill in children an appetite for understanding deep origins and histories. Maybe. When I despair at the state of things, rocks always offer some comfort. I see gneisses and limestones and granites, greenstones and blueschists and red beds, and I think to myself, what a wonderful world.

———

One dreary, snowless Saturday in December, my children and I took a short drive to Green Bay across the remarkably planar bed of Glacial Lake Oshkosh (which is now accreting a stratum of flat-topped warehouses, sprawling motels, and squat fast-food places). In the county hockey arena—just across the street from the hallowed ground of the Packers' Lambeau Field—a remarkable event was taking place. All eleven Native American nations in Wisconsin were gathered for a powwow to celebrate the end of a thirty-year struggle to prevent the opening of a mine near the headwaters of the Wolf River. In its upper reaches in northern Wisconsin, the Wolf is a wild river with lots of white water. It tumbles over 2-billion-year-old volcanic rocks formed when that part of Wisconsin was a swath of submarine volcanic vents spewing out hot metallic brines. Near the tiny town of Crandon, these rocks contain high-grade sulfide ores of copper and nickel, with smaller amounts of gold and silver. The ore deposit is rich but localized, plunging at a steep angle into the ground. Mining it would therefore require blasting deep shafts and continuous pumping of groundwater. But with an estimated 55 million tons of massive sulfides, the Crandon ore body ranks as a world-class deposit, and options to mine it were purchased by a series of multinational corporations, including Exxon, Rio Algom, and BHP Billiton, the world's largest mining company.

These companies apparently did not anticipate the formidable resistance they would face from an unlikely consortium of people who care about this quiet corner of the north woods: fishermen, resort owners, kayakers, left-wing academics from downstate, and, most importantly, members of the Potawatomi and Chippewa tribes, whose reservations are adjacent to the site of the proposed mine. Groundwater pumping would have changed the hydrology of the Wolf River and its connecting waters, including lakes where the tribes harvest wild rice. Tons of pulverized, sulfur-rich mine tailings would have left the river vulnerable to acid drainage for centuries. For thirty years, the mining companies and their opponents played a strategic chess game through the state legislature and Department of Natural Resources, which oversees the mine permitting process. Finally, on October 28, 2003, the Forest County Potawatomi and Sokaogon Chippewa communities announced that they would buy the 6,000-acre mine site for $16.5 million, using revenues from casinos run by the tribes. The copper would stay in the rocks.

As my children and I approached the gymnasium on that gray December day, the building, an architecturally unremarkable and slightly weary-looking structure, seemed to be humming—literally vibrating from within. When we opened the doors, the sound of the relentless drums and high, wailing song hit us like a wave and washed us inside.

I said to my kids, something has happened here.

GLOSSARY

acid: aqueous solution with a higher concentration of hydrogen (H^+) than hydroxide (OH^-) ions.

advection: transport of material or heat via a moving medium (contrast with **diffusion** and **conduction**). **Convection** is a special type of advection.

albedo: reflectivity of a surface; percentage of incoming light or heat energy reflected by an object.

ammonites: ancient marine animals related to modern squid. Their detailed shell ornamentation makes them superb **index fossils** for the Jurassic and Cretaceous periods.

amphibole: a large group of silicate minerals that contain water as an integral part of their crystal structure. Many amphiboles occur as rod or needlelike crystals, and some types form what is commonly called asbestos.

angle of repose: steepest slope that unconsolidated sediment can maintain.

anisotropy: directional variation in the magnitude of a quantity.

anorthite: calcium feldspar, one of the most common minerals in rocks on the Earth and Moon.

aquifer: permeable layer of sediment or rock that carries groundwater.

arc magma: silica (SiO_2)–rich magma produced when water from a subducting ocean slab lowers the melting temperature of the surrounding mantle. This type of magma is typically erupted explosively (as in Japan, Indonesia, and the Cascades). The production of arc magma is unique to Earth and is part of the distillation of continental crust from an interior that is much lower in silica.

asteroid: one of the thousands of rocky bodies orbiting mainly between the inner and outer planets, ranging in size from centimeters to about 1,000 kilometers.

asthenosphere: zone of weak and flowing (but solid) rock in Earth's upper mantle.

atomic mass number: the sum of the number of protons and neutrons in the nucleus of an atom.

banded iron formation: chemical sedimentary rock consisting of centimeter-scale layers of **chert** and iron oxide minerals, including hematite (Fe_2O_3) and magnetite (Fe_3O_4).

basalt: low-silica (**mafic**) volcanic rock, consisting mainly of calcium- and sodium-rich feldspar and pyroxenes. Typical of divergent plate boundaries and oceanic hot spots (and also the lunar lowlands). Intrusive equivalent: gabbro.

bedding: depositional layering in a sedimentary rock.

biogeochemistry: study of the movement and exchange of elements and compounds (e.g., carbon and water) through the biosphere, atmosphere, hydrosphere, and solid earth.

blueschist: metamorphic rock formed by high pressure but relatively low-temperature metamorphism of oceanic crust; a signature of an ancient subduction zone.

carbonate rock: rock consisting of the minerals calcite ($CaCO_3$) or dolomite [$CaMg(CO_3)_2$] (i.e., limestone or dolostone).

catastrophism: pre-nineteenth-century doctrine that Earth was shaped by violent and cataclysmic processes that no longer occur.

chemical sediment: sediment formed by the chemical precipitation of minerals dissolved in water.

chert: chemically precipitated silicon dioxide (SiO_2), compositionally identical to quartz but noncrystalline. Alternates with iron-rich layers in **banded iron formation.**

chondrite (or chondritic meteorite): the most primitive type of meteorite, containing tiny particles called chondrules, which are thought to represent clumps of the original material from which the solar system formed. Chondrites are older than the planets and moons and provide information about the physical and chemical processes during the earliest days of the solar system.

clastic sediment: sediment deposited physically by water, wind, or ice.

climate proxy: an indirect but reasonably continuous record of past climate variation (e.g., tree ring width or the physical and chemical properties of layering in glacial ice).

conduction (thermal): heat transport in a static medium, in which heat energy travels from areas of high temperature to low temperature.

conglomerate: rock composed of pebbles and larger grains (larger than 2 millimeters) deposited by fast-flowing water.

continental shelf: area of shallowly submerged continental crust bordering the continents.

convection (thermal): gravitationally driven overturn of a heated material. The thermal convection of Earth's solid mantle drives **plate tectonics.**

convergent evolution: development of similar anatomical traits in separate lineages of organisms.

convergent plate boundary: boundary at which plates move together via **subduction.**

coprolite: fossilized dung.

core: innermost and densest part of a planet. On Earth, the metallic core begins at a depth of 2,900 kilometers.

correlation: the process of relating rock layers in one area to rocks of the same age in another area.

cosmic ray: high-energy particle that comes from the Sun or from sources outside the solar system and that interacts with atoms in Earth's uppermost atmosphere. Earth's magnetic field shields the surface of the planet from most cosmic rays.

cosmogenic isotope: unstable (radioactive) isotope produced by cosmic ray bombardment of atoms in Earth's uppermost atmosphere. Such isotopes (including ^{12}C and ^{10}Be) are useful for determining how much time has elapsed since an organism or a mineral surface has interacted with the atmosphere.

cross-cutting relationships (principle of): one of several logical principles for determining the sequence of geologic events: If feature A is cut or modified by feature B, then A must be older. See also principles of *inclusion* and *superposition.*

cross section: schematic drawing of a vertical cut into the subsurface.

crust: outermost layer of Earth, 5 to 90 kilometers thick.

crystal: solid in which atoms are arranged in an orderly, spatially repeating three-dimensional array, or lattice.

cyanobacteria: blue-green "algae," single-celled prokaryotic organisms capable of photosynthesis; among the first organisms to appear in the fossil record.

Daisyworld: Earth-like model planet proposed by James Lovelock to explain how his **Gaia hypothesis** is compatible with evolution by natural selection. The model's biosphere consists only of daisies that can influence planetary temperature through their color-related **albedo** (reflectivity).

delta: sediment deposited at the mouth of a river as it enters a standing body of water. Often triangular in shape, the deposit resembles the Greek letter delta.

dendritic: branching or treelike in form.

density: mass of an object per unit volume.

desiccation cracks (mud cracks): polygonal cracks formed on the surface of fine-grained sediment as it dries and contracts.

diamictite: rock composed of two distinct sizes of clastic grains; diamictites usually represent glacial deposits or mudflows.

diffusion: movement of atoms through a static medium, from regions of high concentration to low. Typically very slow in nonmoving liquids (e.g., subsurface magma) and slower still in solids (e.g., mineral lattices).

dislocation: error in the way that atoms in a crystal lattice are arranged; dislocations cause local stress concentration in crystals. These areas significantly reduce mineral strength.

divergent plate boundary: site at which plates move apart. In an ocean basin, a midocean ridge.

ductile flow: slow, viscous, nonbrittle deformation of solid rock.

eclogite: rock formed by high-pressure, moderate-temperature metamorphism of ocean crust. With **blueschist**, eclogite is diagnostic of ancient **subduction** zones.

Ediacaran biota: the first group of organisms to emerge after the **Snowball Earth** interval. First described in the Edicaran Hills of Australia, fossils of these creatures have now been found around the world, but their taxonomic affinities remain controversial. Also called Vendozoans or Vendobionts.

endosymbiosis: a type of **symbiosis** in which two independently evolved organisms (generally one-celled) merge to become a codependent system and eventually a single organism.

entropy: randomness or disorder (the way of the universe).

epicenter: point on the surface of the earth above the **focus** of an earthquake.

eukaryote: an organism whose cells have nuclei.

evaporite: rock composed of salts including halite, gypsum, and anhydrite; formed by the evaporation of saltwater.

exobiologists: scientists concerned with the search for life outside the Earth.

fault: rock fracture along which slip has occurred.

feedback mechanism: a process in which the output (effect) of one phase becomes the input (cause) of the next.

felsic rock: igneous rock high in silica (e.g., **granite**) that could not be derived in a single melting episode from the Earth's mantle. Contrast **mafic rock**.

fluid: any substance that flows (may be solid, liquid, or gas).

flux: an element or a compound that lowers the melting temperature of a pure substance.

focus (earthquake): the origin of an earthquake; point in the subsurface where fault slip begins. Also called the **hypocenter**.

fold: wrinkle or buckle in layered rock, usually formed in response to layer-parallel shortening.

food chain or web: a hierarchy of organisms in which energy and nutrients are transferred from primary producers to the organisms that consume them and then to the organisms that consume those organisms, and so on.

fossil: preserved remains or other traces of a living organism.

fossil fuels: hydrocarbons (compounds of hydrogen and carbon) formed from incompletely decayed organic matter that can be burned as fuels; include coal, oil, and natural gas.

fractal: geometric feature that looks the same over a range of scales.

fractional melting: stepwise melting of rock beginning with minerals that have the lowest melting temperature.

Gaia hypothesis: view that Earth's near-surface environment has been largely controlled by biological activity ,i.e. that organisms have acted to minimize fluctuations in surface temperatures and air and water chemistry.

galena: lead sulfide (PbS); principal **ore** of lead.

gas hydrate: frozen form of methane and other biogenic gases, found in many parts of the world in the upper layers of sediment on the ocean floor. These frozen hydrocarbons may volatilize suddenly if pressures or temperatures change.

geobarometer: mineral or group of minerals indicative of the peak pressures experienced by a metamorphic rock.

geomagnetic time scale: time scale based on the history of reversals in the polarity of Earth's magnetic field. Absolute ages have been assigned to the reversals through isotopic dating of rocks formed close to the time of a reversal.

geomicrobiology: scientific subdiscipline that focuses on understanding the relationships between microorganisms and minerals.

geophysiology: the study of the hydrosphere, atmosphere, biosphere, and solid Earth as a self-regulating system of interconnected subsystems (a scientifically palatable name for the **Gaia hypothesis**).

geothermal gradient: increase in temperature with depth in Earth's subsurface, expressed as °C per kilometer.

geothermometer: mineral or group of minerals indicative of the peak temperatures experienced by a metamorphic rock.

gneiss: intensely metamorphosed rock with prominent mineral banding.

gradualism: ultra-uniformitarian doctrine that the Earth has been shaped only by slow, gradual processes (a reaction against **catastrophism**).

granite: intrusive igneous rock, typical of continental crust. Consists principally of sodium or potassium feldspar, quartz, hornblende, and mica.

gravitational differentiation: settling of materials according to density.

greenhouse effect: trapping of heat close to a planet's surface by gases that transmit incoming sunlight but block heat reflected back from the planet's surface. Carbon dioxide (CO_2), methane (CH_4), and water vapor (H_2O) can all act as greenhouse gases.

greenstone: basalt that has been subjected to hydrothermal alteration (low-grade metamorphism in the presence of heated groundwater).

groundwater: water contained in the pore spaces of soil, sediment, and rock.

half-life: time required for half the amount of an unstable parent isotope to decay to a daughter isotope.

highlands (lunar): older, heavily cratered terrain on the Moon.

histogram: plot showing the frequency with which particular values of a variable occur.

hot spot: point on Earth's surface above an isolated column of unusually hot, buoyantly rising mantle rock. Magma generated as this rock nears the surface causes volcanic activity like that in Hawaii.

hydrologic cycle: the continuous cycling of water in different forms at, above, and under the Earth's surface.

hypocenter: the origin of an earthquake; point in the subsurface where fault slip begins. Also called the **focus**.

hypsometry: mathematical description of the topography of a planet.

igneous rock: rock formed from the molten state (magma).

inclusion (principle of): principle that a rock containing pieces of other rocks must be younger than those pieces (or conversely, that the pieces must be older than the rock that contains them).

incompatible element: element that does not fit readily into mantle minerals, typically because of a large atomic radius.

index fossil: fossil of an organism whose species existed for a geologically brief period. Index fossils are very useful in **correlation**.

index mineral: metamorphic mineral formed only under restricted pressure or temperature conditions and thus diagnostic of a particular depth or thermal setting.

intrusive rock: igneous rock emplaced into the subsurface.

iridium: element 77, very rare on Earth, but found in higher concentrations in meteorites. High iridium concentrations in rocks from the Cretaceous-Tertiary boundary first led to speculations that a meteorite impact may have led to the extinction of the dinosaurs.

isotope: atom of an element that can have varying numbers of neutrons in its nucleus. Isotopes may be either stable or prone to decay (radioactive).

isotopic age: time elapsed since the crystallization of a mineral or rock as determined from the ratio of daughter to parent isotopes.

Lagerstätten: fossil bed in which the organisms, especially soft-bodied ones, are preserved with unusual anatomical detail (from German word meaning "lode place").

land ethic: conservationist Aldo Leopold's phrase for a sense of moral responsibility toward not only humans but also land, water, and all of the biosphere.

limestone: rock composed mainly of calcite; typically precipitated from seawater.

limiting nutrient: element or molecule essential for life but available in limited quantities; if more of the nutrient were available, an organism's growth would be enhanced.

lithosphere: strong outer layer of the solid Earth, including the **crust** and uppermost **mantle**. Typically about 100 kilometers thick.

logarithmic scale: measurement scale in which each unit represents an order-of-magnitude difference in value. The Richter scale is a logarithmic scale.

lowlands (lunar): smooth, lava-covered terrain on the Moon. Younger than the lunar highlands but still ancient by Earth standards.

mafic rock: magnesium-rich, comparatively silicon-poor igneous rock that could be derived from Earth's mantle in one stage of **fractional melting**. **Basalt** is a mafic volcanic rock.

magnitude (earthquake): quantitative description of the size of an earthquake, based on the amplitude of recorded peaks on seismic records.

mantle: layer surrounding the core of each of the inner planets of the solar system. Earth's rocky mantle constitutes 83 percent of the planet's volume and extends from about 40 to 2,900 kilometers below the surface.

marble: metamorphosed, recrystallized limestone or dolomite.

mass extinction: abrupt disappearance of many species from the fossil record.

metamorphic rock: rock modified by temperature, pressure, or chemical conditions different from those at which it formed.

metamorphism: growth of new minerals in rocks in response to physical or chemical conditions different from those at which they formed.

meteorite: an extraterrestrial rock that has fallen to Earth. Most meteorites are thought to have come from the asteroid belt, although a few are impact ejecta from other planets (especially Mars).

mineral: a natural crystalline solid (one in which atoms are arranged in regular, three-dimensional arrays) with a well-defined chemical composition. Most rocks consist of more than one type of mineral.

mudstone: see shale.

mutation (genetic): a change in the genetic code (DNA) within a cell. Some mutations are harmless; others can lead to cancer and birth defects.

natural selection: evolution of organisms by the selective survival of the fittest.

negative feedback mechanism: a self-correcting process.

nitrogen fixation: biochemical process in which bacteria convert atmospheric nitrogen (N_2) to compounds including ammonium (NH_4^+).

noble gas: any of the gaseous elements that occur in the last column of the periodic table. Having full outer electron shells, they do not bond with other elements.

nucleosynthesis: fusion of elements in the cores of large stars to form heavier elements.

obsidian: volcanic glass.

ore: economically viable source of a metallic element.

orogen: ancient mountain belt, typically one that is deeply eroded and has little topographic expression but is marked by deformed and metamorphosed rocks.

oxidation (of an element or a compound): ionization by the loss of electrons, usually to oxygen atoms.

ozone: trivalent oxygen (O_3). Naturally occurring ozone in the Earth's upper atmosphere (stratosphere) shields the Earth's surface from much of the ultraviolet light from the Sun. Ozone emitted at ground level by internal-combustion engines is a health risk and pollutant.

paleosol: clay-rich sedimentary rock representing an ancient soil horizon.

palimpsest: a sedimentary bed that was deposited during one event and then reworked in a later event. More generally, any feature that preserves a partly overwritten record of multiple stages or epochs.

Pangaea: supercontinent assembled by plate collisions in late Paleozoic time (ca. 300–270 million years ago), which built the Appalachian, Caledonian, and Hercynian mountain belts. Pangaea broke up as the modern Atlantic Ocean opened.

peridotite: igneous rock composed principally of the mineral olivine (peridot). Rock type characteristic of Earth's upper mantle.

permeability: measure of the ease with which fluids can move through sediment or rock.

photosynthesis: process by which green plants, with sunlight as their energy source, convert water and atmospheric carbon dioxide to carbohydrates and free oxygen.

plate tectonics: theory that Earth's **lithosphere** is a mosaic of moving pieces, or plates.

polarity (magnetic): the sense of magnetization of a body—that is, the relative positions of its north and south poles.

positive feedback mechanism: a self-perpetuating process.

postglacial rebound: slow rise of land formerly covered by a thick ice sheet. Occurs as ductile mantle rock that was displaced because of the weight of the ice gradually flows back into place.

preservation potential: the likelihood that a sedimentary deposit, fossil, or other artifact of a particular period will be preserved for some time into the future.

primary producers: organisms (e.g., photosynthesizing plants and chemosynthetic bacteria) able to derive energy from inorganic sources. Also called *autotrophs*.

prokaryotes: single-celled organisms, including bacteria and some algae, that lack nuclei.

protolith: original rock from which a **metamorphic** rock formed.

pumice: rock formed from gas-rich, glassy volcanic "foam." Pumice can contain so much air space that it will float.

punctuated equilibrium: theory that evolution over geologic time does not occur at a constant rate but is instead characterized by periods of relative stasis alternating with intervals of rapid innovation (often in response to major environmental changes).

radioactive decay, or **radioactivity:** spontaneous breakdown of unstable **isotopes** into heat, subatomic particles, and other isotopes.

rare-earth element: any of the elements with atomic numbers from 57 to 71 (lanthanum through lutetium). Although they are present in only minute quantities in rocks, the relative abundances of rare earth elements can help pinpoint the source of magma that formed an igneous rock.

red beds: clastic sedimentary rock deposited on land and colored red-orange by oxidized iron.

reduction (of an element or a compound): ionization by the addition of electrons.

reservoir: temporary storage place for a particular material.

residence time: average length of time a material remains in a particular setting.

rhyolite: high-silica (silicic) volcanic rock consisting mainly of potassium- and sodium-rich feldspar, quartz, and mica. Common at **convergent plate boundaries** and continental hot spots. Intrusive equivalent: granite.

rhythmite: sedimentary deposit that accumulated in response to annual or seasonal cycles (e.g., **varves** in glacial lakes).

Richter scale: most commonly used earthquake magnitude scale, based on the amplitudes of recorded peaks on a seismogram.

rock: natural aggregate of minerals. Most rocks consist of more than one type of mineral.

Rodinia: supercontinent that existed in latest Proterozoic time (ca. 750–600 million years ago, a tectonic generation older than **Pangaea**). Rodinia's low-latitude position may have been responsible for the **Snowball Earth** ice age.

sandstone: sedimentary rock composed of sand-sized grains ($^1/_{16}$–2 millimeters diameter).

scaling law: mathematical description of the relationship between population sizes at different levels within a hierarchical system such as a food web.

schist: metamorphic rock formed from fine-grained sedimentary rocks (mudstones).

seafloor spreading: process in which new oceanic lithosphere is formed by volcanic activity at midocean ridges and then displaced progressively outward.

sediment fan: fan-shaped accumulation of river sediments at the base of a steep slope. If located at the base of the continental shelf, termed a *submarine fan*; at the foot of a mountain on land, an *alluvial fan*.

sedimentary rock: rock composed of recycled pieces of earlier rocks, deposited by water, wind, or ice.

sedimentary structures: physical features in sediments and sedimentary rocks that reflect their mode of deposition.

seismic wave: vibrational wave triggered by sudden slip on a **fault.** The arrival of seismic waves at the surface of the Earth is what causes the ground motion in an earthquake.

seismogram: visual or digital record of ground motion during an earthquake, generated by a motion-detecting instrument called a seismometer.

shale: sedimentary rock composed of clay-sized grains (microscopic particles smaller than 1/256 of a millimeter); deposited in quiet bodies of water. Also called *mudstone.*

shield: core of a continent, where the oldest rocks are found.

silicate: the category to which more than 95 percent of the minerals on Earth belong with silicon and oxygen as the fundamental constituents.

slip (seismic): the amount of offset or displacement across a fault in an earthquake event; typically less than 1 meter.

Snowball Earth hypothesis: proposal that Earth experienced as many as four periods of extreme cold (perhaps cold enough to freeze the oceans) during latest Proterozoic time (ca. 750–600 million years ago).

solar nebula: cloud of interstellar gas thought to have collapsed under its own gravity to form the Sun and planets.

stratigraphy: the study of layered rock sequences and what they record about the changes at the surface of the Earth over time.

stress: unequal forces applied in different directions; causes rocks to deform.

stromatolite: finely layered sedimentary rock formed by mats of single-celled organisms, often blue-green algae, that live on tidal flats. First appear in the rock record in the mid-Precambrian.

subduction: sinking of old, cold, and dense ocean crust into the mantle at deep ocean trenches.

supernova: explosive death of a large star at the end of its life.

superposition (principle of): principle that in a layered sequence of rocks accumulated at the Earth's surface, the oldest rocks are at the bottom.

sverdrup: unit of volumetric flow rate used to quantify the magnitude of ocean currents. Equal to 1 million cubic meters per second (m^3/second).

symbiosis: mutually beneficial cooperative living arrangement between two distinct organisms.

taphonomy: the study of the factors that contribute to the preservation and fossilization of organic remains.

tectonic plate: one of about a dozen moving pieces that make up Earth's outermost layer, the **lithosphere.**

tectonics: see *plate tectonics.*

teleological: implying purposeful design.

tempestite: sedimentary rock deposited in a violent storm.

terrestrial sediments: sediments deposited in land environments.

thermal conduction: see *conduction (thermal).*

thermal convection: see *convection (thermal).*

thermodynamics: branch of physics that studies the relationship between heat and other forms of energy.

thermohaline ocean circulation: convective motion of ocean currents around the globe, driven by the differences in water density associated with temperature and salinity variations in seawater.

trace fossil: track, burrow, or other marks left by a living organism in sediment.

trench: deep ocean trough at the site of **subduction.**

trilobites: early arthropods (ancestors of modern crustaceans); excellent Cambrian and Ordovician **index fossils.**

trophic level: tier within a food chain or web (e.g., primary producer, herbivore, insectivore, carnivore).

turbidite: sedimentary rock laid down on the deep seafloor by a turbidity current (a submarine landslide).

unconformity: within a rock sequence, a surface that represents a period of erosion (or nondeposition); a gap in the rock record.

uniformitarianism: principle that the same processes that operate on Earth at present also operated in the geologic past.

varve: a pair of sedimentary layers deposited in a glacial lake or fjord representing a single year's accumulation; typically a sandy layer recording summertime transport of sediment by flowing streams, and a clayey layer formed when fine sediment settled out beneath the frozen water surface in winter.

vein: mineral-filled rock fracture. In some veins, the mineral fillings formed from magma; in others, they were precipitated from groundwater.

viscosity: the degree to which a fluid resists flow.

Younger Dryas: a brief return to full glacial conditions near the end of the last Ice Age, about 11,000 years ago.

zircon: a mineral consisting of zirconium, silicon, and oxygen ($ZrSiO_4$) and trace amounts of uranium. Because of its hardness and high melting temperature, zircon can remain unaltered through many cycles of erosion and metamorphism and is therefore ideal for **isotopic age** dating.

Notes

Prologue

1. I have provided a glossary at the end of this book to help ease the reader's path toward geological fluency. These glossary terms appear in boldface the first time they occur in the book.

2. Here and throughout the book, I use *billion* in the American sense, meaning one thousand million.

3. S. Wilde et al., "Evidence from Detrital Zircons for the Existence of Continental Crust and Oceans on the Earth 4.4 Gyr Ago," *Nature* 409 (2001): 175–178.

Chapter 1

1. Charles Darwin, *The Origin of Species* (London: Penguin Books, 1985; first published by John Murray, 1859), 297.

2. Joe Burchfield, *Lord Kelvin and the Age of the Earth* (Chicago: University of Chicago Press, 1990).

3. In 1996, a group of NASA scientists reported indirect evidence for ancient microorganisms within a 1.3-billion-year-old meteorite believed to be of martian origin (David S. McKay et al., "Search for Past Life on Mars: Possible Relic Biogenic Activity in Martian Meteorite ALH84001," *Science* 273 [1996]: 924–930). Most scientists are skeptical of this interpretation, however (e.g., Ralph Harvey and Harry McSween Jr., "A Possible High-Temperature Origin for the Carbonates in the Martian Meteorite ALH84001," *Science* 273

[1996]: 757–762). Even those who contend that the meteorite contains fossils of an ancient martian life-form concur that today the planet is lifeless.

4. Quoted in Connie Barlow, ed., *From Gaia to Selfish Genes* (Cambridge, Mass.: MIT Press, 1991), 2.

5. James Lovelock, "Gaia As Seen Through the Atmosphere," *Atmospheric Environment* 6 (1972): 579–580; James Lovelock and Lynn Margulis, "Atmospheric Homeostasis by and for the Biosphere," *Tellus* 26 (1974): 2–9; Lynn Margulis and James Lovelock, "Biological Modulation of the Earth's Atmosphere," *Icarus* 21 (1974): 471–489.

6. A. J. Watson and James Lovelock, "Biological Homeostasis of the Global Environment," *Tellus* 35B (1983): 284–289. Also James Lovelock, *The Ages of Gaia: A Biography of Our Living Earth* (New York: W. W. Norton, 1988), 35–64.

CHAPTER 2

1. Stephen Baxter, *Revolutions in the Earth: James Hutton and the True Age of the Earth* (London: Weidenfeld and Nicholson, 2003).

2. Stephen Jay Gould, *Time's Arrow, Time's Cycle: Myth and Metaphor in the Discovery of Geologic Time* (Cambridge, Mass.: Harvard University Press, 1988), 66–80.

3. James Hutton, *Theory of the Earth*, Philosophical Transactions, Royal Society of Edinburgh I, part II (1788): 209–304.

4. Gould, *Time's Arrow*, 146–147.

5. Mark Twain, *Autobiography*, ed. Charles Neider (New York: Harper and Brothers, 1959), 83. I find it amazing that Lyell's *Principles of Geology* was in circulation, or at least known outside scientific circles, in the hinterlands of North America in the 1850s.

6. S. Wilde et al., "Evidence from Detrital Zircons for the Existence of Continental Crust and Oceans on the Earth 4.4 Gyr Ago," *Nature* 409 (2001): 175–178.

7. Derek Ager, *The Nature of the Stratigraphical Record* (London: Macmillan, 1973), 43–50.

8. C. P. Sonett et al., "Late Proterozoic and Paleozoic Tides, Retreat of the Moon, and Rotation of the Earth," *Science* 273 (1996): 100–104.

9. See the recent Hutton biography by Jack Repcheck, *The Man Who Found Time: James Hutton and the Discovery of Earth's Antiquity* (New York: Perseus, 2003).

10. Stephen Jay Gould, in his best-seller *Wonderful Life: The Burgess Shale and the Nature of History* (New York: Norton, 1988), brought the Burgess Shale to the attention of nongeologists. He emphasized the anatomical diversity of the Burgess creatures, supporting his view that evolution proceeds in fits and starts. Some of his interpretations have been refuted by Derek Briggs, *The Fossils of the Burgess Shale* (Washington, D.C.: Smithsonian Press, 1995), and especially Simon Conway-Morris, *The Crucible of Creation: The Burgess Shale and the Rise of Animals* (Los Angeles: Getty Center, 1999), in unusually polemical prose.

11. See Simon Winchester, *The Map That Changed the World: William Smith and the Birth of Modern Geology* (New York: Harper-Collins, 2001).

12. S. A. Bowring et al., "Calibrating Rates of Early Cambrian Evolution," *Science* 261 (1993): 1293–1298.

13. S. A. Bowring and T. Housh, "The Earth's Early Evolution," *Science* 269 (1995): 1535–1540.

14. Clair Patterson, "Age of Meteorites and the Earth," *Geochimica et Cosmochimica Acta* 10 (1956): 230–237.

CHAPTER 3

1. Dan Krotz, "Another Magnet, Another Record," *Science Beat*, Lawrence Berkeley Laboratory, 9 January 2004, www.lbl.gov/Science-Articles/Archive/sb-AFRD-magnet-record.html (15 July 2004).

2. Charles Darwin, *Journal of Researches into the Natural History and Geology of the Countries Visited During the Voyage of H.M.S.* Beagle *Round the World* (New York: Harper and Brothers, 1859), 2:45–59.

3. For a riveting account of this disastrous unriveting and a good overview of the theory of fracturing, see Mark Eberhart, "Why Things Break," *Scientific American* 155 (1999): 66–73.

4. F. Vine and D. Matthews, "Magnetic Anomalies Over Oceanic Ridges," *Nature* 199 (1963): 947–949.

5. G. Glatzmaier and P. Roberts, "A Three-Dimensional Self-Consistent Computer Simulation of Geomagnetic Field Reversal," *Nature* 377 (1995): 203–209.

6. R. Coe and M. Prevot, "Evidence Suggesting Extremely Rapid Field Variation During a Geomagnetic Reversal," *Earth and Planetary Science Letters* 92 (1989): 292–298.

7. The Antarctic ice cap is considerably older than that in Greenland, and the longest continuous core thus far recovered from it dates back 740,000 years. That core is described in European Project for Ice Coring in Antarctica (EPICA) community members, "Eight Glacial Cycles from an Antarctic Ice Core," *Nature* 429 (2004): 623–628.

8. Richard Alley, *The Two-Mile Time Machine: Ice Cores, Abrupt Climate Change and Our Future* (Princeton, N.J.: Princeton University Press, 2001), 126.

9. Benoit Mandelbrot, "How Long Is the Coast of Britain? Statistical Self-Similarity and Fractal Dimension," *Science* 155 (1967): 636–638.

10. Jonathan Swift, *Poems*, ed. Harold Williams (Oxford: Clarendon Press, 1937).

11. Lewis Richardson, "The Supply of Energy from and to Atmospheric Eddies," *Proceedings of the Royal Society of London, Series A* 97 (1920): 354–373.

12. A. Belgrano et al., "Allometric Scaling of Maximum Population Density: A Common Rule for Marine Phytoplankton and Terrestrial Plants," *Ecology Letters* 5 (2002): 611–613. Also G. B. West, J. Brown, and B. Enquist, "A General Model for the Origin of Allometric Scaling Laws in Biology," *Science* 276 (1997): 122–126.

13. R. J. Parkes et al., "Deep Bacterial Biosphere in Pacific Ocean Sediments," *Nature* 371 (1994): 410–413.

14. Dorion Sagan and Lynn Margulis, *Garden of Microbial Delights: A Practical Guide to the Subvisible World* (Orlando, Fl.: Harcourt Brace Jovanovich, 1988).

15. John Alroy, Charles Marshall, and Arnie Miller, *The Paleontology Database Project*, 22 August 2000, www.paleodb.org (15 July 2004). Also J. Alroy et al., "Effects of Sampling Standardization on Estimates of Phanerozoic Marine Diversification," *Proceedings of the National Academy of Sciences* 98 (2001): 6261–6266.

16. For laboratory models of ecosystems, see O. Petchey, P. McPhearson, and T. Casey, "Environmental Warming Alters Food-Web Structure and Ecosystem Function," *Nature* 402 (1999): 69–72.

17. Jorge Luis Borges, *Collected Fictions*, trans. Andrew Hurley (New York: Penguin, 1999), 289–328.

18. Evelyn Fox Keller, *A Feeling for the Organism: The Life and Work of Barbara McClintock* (San Francisco: W. H. Freeman, 1983).

CHAPTER 4

1. John Gribbin, *Stardust: Supernovae and Life, the Cosmic Connection* (New Haven, Conn.: Yale University Press, 2001), 156.

2. T. Lee, D. A. Papanastassiou, and G. J. Wasserburg, "Demonstration of ^{26}Mg Excess in Allende and Evidence for ^{26}Al," *Geophysical Research Letters* 3 (1976): 109–112.

3. The idea was first proposed by W. K. Hartmann and D. R. Davis, "Satellite-Sized Planetesimals and Lunar Origin," *Icarus* 24 (1975): 504–515. More recent supercomputer simulations have demonstrated the physical plausibility of the giant impact hypothesis, e.g., R. M. Canup and L. W. Esposito, "Accretion of the Moon from an Impact-Generated Disk," *Icarus* 119 (1996): 427–446.

4. The first half of the comet's name acknowledges the pioneering planetary geologist Eugene Shoemaker and his wife, Caroline, whose decades of research at the U.S. Geological Survey helped open geologists' eyes to the importance of impact cratering on Earth. Eugene died tragically in a car accident in 1997, and the following year, his ashes were carried to the Moon on NASA's *Lunar Prospector* spacecraft. When *Prospector's* mission ended in 1999, the craft was deliberately crashed on the lunar surface. In this way, Shoemaker realized posthumously his dream of landing on the Moon. In an eerie way, Shoemaker's journey mirrors that of the comet that carried his name.

5. For a good, nontechnical summary of current thinking on this topic, see Ben Harder, "Water for the Rock: Did Earth's Oceans Come from the Heavens? Research Into the Origin of the Earth's Seas," *Science News*, 23 March 2002.

6. Not all geologists agree about this; some prefer the more uniformitarian view that plate tectonics began very early in Earth's history. But if we look to Venus and Mars for clues about earlier stages in Earth's evolution, we see no evidence of well-defined plate boundaries, subduction zones, or differentiation into two crustal types. This, together with the thermal arguments discussed in the text, makes it seem unlikely that Earth's plate tectonic system operated in its modern form in Archean time.

7. Paul F. Hoffman, "Wopmay Orogen: A Wilson Cycle of Early Proterozoic Age in the Northwest of the Canadian Shield," in *The Continental Crust and Its Mineral Deposits*, ed. D. W. Strangway, Geological Association of Canada Special Paper 20 (1980): 523–549; A. Möller et al., "Evidence of a 2 Ga Subduction Zone: Eclogites in the Usagaran Belt of Tanzania," *Geology* 23 (195): 1067–1070.

8. S. Bowring and T. Housh, "The Earth's Early Evolution," *Science* 269 (1995): 1535–1540. See also R. Rudnick, "Making Continental Crust," *Nature* 378 (1995): 571–578.

9. S. Wilde et al., "Evidence from Detrital Zircons for the Existence of Continental Crust and Oceans on the Earth 4.4 Gyr Ago," *Nature* 409 (2001): 175–178.

10. This has been documented in my own work with Håkon Austrheim (University of Oslo) and his students on some deep crustal rocks north of Bergen in western Norway. M. Bjornerud, H. Austrheim, and M. Lund, "Processes Leading to Densification (Eclogitization) of Tectonically Buried Crust," *Journal of Geophysical Research* 107 (B10) (2002), DOI 10.1029/2001JB000527; and M. Bjornerud and H. Austrheim, "Inhibited Eclogite Formation: The Key to the Rapid Growth of Strong and Buoyant Archean Crust," *Geology* 32 (2004): 765–768.

11. Paradoxically, erosion can sometimes cause mountains to *grow*, as a result of the unloading of the underlying mantle as rock is removed from the surface. This phenomenon, called *isostatic rebound*, is very similar to the slow but measurable uplift of land that was previously covered by ice (Chapter 3). Whether erosion causes net reduction in topographic elevations depends of the relative rates of sediment removal (a function of climate) and mantle flow.

12. The classic analysis of the Grand Banks turbidity current is P. Kuenen, "Estimated Size of the Grand Banks Turbidity Current," *American Journal of Science* 250 (1952): 874–884.

13. D. Rothman, J. Grotzinger, and P. Fleming, "Scaling in Turbidite Deposition," *Journal of Sedimentary Research* A64 (1994): 59–67.

14. S. R. Taylor and S. M. McLennan, "The Geochemical Evolution of the Continental Crust," *Reviews of Geophysics* 33 (1995): 241–265.

15. F. Tera et al., "Sediment Incorporation in Island-Arc Magmas: Inferences from [10]Be," *Geochimica et Cosmochimica Acta* 50 (1986): 535–550. See also J. D. Morris, W. P. Leeman, and F. Tera, "The Subducted Component in Island Arc Lavas: Constraints from Be Isotopes and B-Be Systematics," *Nature* 344 (1990): 31–36.

16. An engaging, nontechnical account of the Snowball Earth hypothesis is Gabrielle Walker, *Snowball Earth: The Story of the Great Global Catastrophe That Spawned Life as We Know It* (New York: Crown Publishers, 2003).

17. J. Kirschvink, "Late Proterozoic Low-Latitude Glaciation: The Snowball Earth," in *The Proterozoic Biosphere,* ed. W. Schopf, C. Klein, and D. DesMaris (Cambridge: Cambridge University Press, 1992), 51–52. P. Hoffman et al., "A Neoproterozoic Snowball Earth," *Science* 281 (1998): 1342–1346.

18. M. Budyko, "The Effect of Solar Radiation Variations on the Climate of the Earth," *Tellus* 21 (1969): 611–619.

19. J. Kirschvink, R. L. Ripperdan, and D. A. Evans, "Evidence for a Large-Scale Reorganization of Early Cambrian Continental Masses by Inertial Interchange True Polar Wander," *Science* 277 (1997): 541.

20. R. Pierrehumbert, "High Levels of Carbon Dioxide Necessary for the Termination of Global Glaciation," *Nature* 429 (2004): 646–649.

21. M. Kennedy, N. Christie-Blick, and L. Sohl, "Are Proterozoic Cap Carbonates and Isotopic Excursions a Record of Gas Hydrate Destabilization Following Earth's Coldest Intervals?" *Geology* 29 (2001): 443–446.

22. Two excellent books on the end-Permian extinction, both written by paleontologists for nonspecialist readers, are Michael Benton, *When Life Nearly Died* (London: Thames and Hudson, 2003), and Douglas Erwin, *The Great Paleozoic Crisis: Life and Death in the Permian* (New York: Columbia University Press, 1993).

23. S. Bowring et al., "U/Pb Zircon Geochronology and Tempo of the End-Permian Mass Extinction," *Science* 280 (1998): 1039–1045.

24. M. K. Reichow et al., "[40]Ar/[39]Ar Dates on Basalts from the West Siberian Basin: Doubled Extent of the Siberian Flood Basalt Province," *Science* 296 (2002): 1846–1849.

25. R. Twitchett and P. Wignall, "Ocean Anoxia and the End-Permian Mass Extinction," *Science* 272 (1996): 1155–1158.

26. E. S. Krull and G. J. Retallack, "$\delta^{13}C$ Depth Profiles from Paleosols Across the Permian-Triassic Boundary: Evidence for Methane Release," *Geological Society of America Bulletin* 112 (2000): 1459–1472.

27. R. Twitchett et al., "Rapid and Synchronous Collapse of Marine and Terrestrial Ecosystems During the End-Permian Biotic Crisis," *Geology* 29 (2001): 351–354.

28. G. Retallack, J. Veevers, and R. Morante, "Global Coal Gap Between Permian-Triassic Extinction and Middle Triassic Recovery of Peat-Forming Plants," *Geological Society of America Bulletin* 108 (1996): 195–207.

29. L. Becker et al., "A Possible End-Permian Impact Crater Offshore of Northwestern Australia," *Science* 304 (2004): 1469–1476.

30. W. Broecker and W. Farrand, "Radiocarbon Age of the Two Creeks Forest Bed, Wisconsinan," *Geological Society of America Bulletin* 74 (1963): 795–802.

31. T. Stocker, D. Wright, and W. Broecker, "The Influence of High-Latitude Surface Forcing on the Global Thermo-Haline Circulation," *Paleoceanography* 7 (1992): 529–541.

32. J. T. Teller, D. W. Leverington, and J. D. Mann, "Freshwater Outbursts to the Oceans from Glacial Lake Agassiz and Global Change During the Last Deglaciation," *Quaternary Science Reviews* 21 (2002): 879–887.

Chapter 5

1. Lynn Margulis, *Symbiosis and Cell Evolution*, 2nd ed. (San Francisco: Freeman, 1992).

2. T. M. Han and B. Runnegar, "Megascopic Eukaryotic Algae from the 2.1-Billion-Year-Old Negaunee Iron-Formation, Michigan," *Science* 257 (1992): 232–235.

3. B. K. Pierson, "The Emergence, Diversification, and Role of Photosynthetic Eubacteria," in *Early Life on Earth, Proceedings of Nobel Symposium 84*, ed. S. Bengston (New York: Columbia University Press, 1995), 161–180.

4. Lynn Margulis and Dorion Sagan, *What Is Life?* (New York: Simon and Schuster, 1995); A. Knoll, "The Early Evolution of Eukaryotes: A Geological Perspective," *Science* 256 (1992): 622–627.

5. J. Maynard Smith, *The Evolution of Sex* (Cambridge: Cambridge University Press, 1978).

6. N. Barton and B. Charlesworth, "Why Sex and Recombination?" *Science* 281 (1998): 1986–1990.

7. J. Felsenstei, "Sex and the Evolution of Recombination," in *The Evolution of Sex: An Examination of Current Ideas,* ed. R. E. Michod and B. R. Levin (Sunderland, Mass.: Sinauer Associates, 1987), 74–86.

8. R. Dunbrack, C. Coffin, and R. Howe, "The Cost of Males and the Paradox of Sex: An Experimental Investigation of the Short-Term Competitive Advantages of Evolution in Sexual Populations," *Proceedings of the Royal Society of London, Series B* 262 (1995): 45–49.

9. Some paleobiologists believe that sexual reproduction may not have been "invented" until later in Proterozoic time. See, for example, N. Butterfield, "*Bangiomorpha pubescens* N. Gen., N. Sp.: Implications for the Evolution of Sex, Multicellularity, and the Mesoproterozoic/Neoproterozoic Radiation in Eukaryotes," *Paleobiology* 26 (2000): 386–404. But given the costliness of sexual reproduction to organisms, it seems more likely that sex arose at a time when strong new environmental pressures began to make the rewards of sex (the capacity for greater evolutionary innovation) greater than its risks.

10. For the complete story of the importance of communal living on Earth, see Lynn Margulis, *Symbiotic Planet: A New Look at Evolution* (New York: Basic Books, 2000).

11. B. Rasmussen et al., "Discoidal Impressions and Trace-Like Fossils More Than 1200 Million Years Old," *Science* 296 (2002): 1112–1115.

12. Butterfield, "*Bangiomorpha pubescens,*" 386.

13. For these organisms' resemblance to fungi and algae, see G. J. Retallack, "Were the Ediacaran Fossils Lichens?" *Paleobiology* 20 (1994): 523–544. For their connection to arthropods and jellyfish, see M. F. Glaessner, *The Dawn of Animal Life* (Cambridge: Cambridge University Press, 1984). For the theory that these organisms have no living descendants, see Adolf Seilacher, "Vendozoa: Organismic Construction in the Proterozoic Biosphere," *Lethaia* 22 (1989): 229–239.

14. M. Clapham and G. Narbonne, "Ediacaran Epifaunal Tiering," *Geology* 30 (2002): 627–630.

15. G. Narbonne and J. Gehling, "Life After Snowball: The Oldest Complex Ediacaran Fossils," *Geology* 31 (2003): 27–30.

16. I borrow this imagery from a controversial book by paleontologist Mark McMenamin, *The Garden of Ediacara* (New York: Columbia University Press, 1998). It provides a fascinating if sometimes fanciful account of the Ediacaran ecosystems.

17. Stephen Jay Gould, *Wonderful Life: The Burgess Shale and the Nature of History* (New York: W. W. Norton, 1990).

18. A pair of articles by Conway Morris and Gould on the Burgess creatures was published as a (very pointed) point-counterpoint in Simon Conway Morris and Stephen Jay Gould, "Showdown on the Burgess Shale," *Natural History,* December 1998, 48–55.

19. Richard Dawkins, *Unweaving the Rainbow: Science, Delusion and the Appetite for Wonder* (New York: Mariner Books, 2000).

20. Richard Dawkins, review of *Wonderful Life,* by Stephen Jay Gould, in *Sunday Telegraph* (London), 25 February 1990.

21. Stephen Jay Gould and Niles Eldredge, "Punctuated Equilibria: The Tempo and Mode of Evolution Reconsidered," *Paleobiology* 3 (1977): 115–151.

22. To Gould, the illogical layout of typewriter keys, now inherited by computer keyboards, is a classic case of accidental survival of a lucky but imperfect design that just happened to emerge at the right time. See Stephen Jay Gould, "The Panda's Thumb of Technology," in *Bully for Brontosaurus* (New York: W. W. Norton, 1991), 59–75.

23. The lottery and tape metaphors are from Gould, *Wonderful Life.*

24. Simon Conway Morris, *The Crucible of Creation* (Oxford: Oxford University Press, 1998).

25. For Ediacaran fauna dates, see J. Grotzinger et al., 1995, "Biostratigraphic and Geochronologic Constraints on Early Animal Evolution," *Science* 270 (1995): 598–604. For Burgess Shale fossil dates, see D.-G. Shu et al., "Lower Cambrian Vertebrates from South China," *Nature* 402 (1999): 42–46.

26. M. Fedonkin and B. Waggoner, "The Late Precambrian Fossil Kimberella Is a Mollusc-Like Bilaterian Organism," *Nature* 388 (1997): 868–871.

27. S. Bengtson, "The Cap-Shaped Cambrian Fossil Maikhanella and the Relationship Between Coeloscleritophorans and Molluscs," *Lethaia* 25 (1992): 401–420.

28. W. F. Lloyd, *Two Lectures on the Checks to Population* (Oxford: Oxford University Press, 1833). The phrase was also used as the title of a now-classic environmental essay by Garrett Hardin, "The Tragedy of the Commons," *Science* 162 (1968): 1243–1248.

29. W. F. Lloyd, quoted in Hardin, "The Tragedy of the Commons."

30. An entire book about the role of vision in the Cambrian explosion is Andrew Parker, *In the Blink of an Eye* (New York: Perseus Books, 2003).

31. A readable overview of the fossil record of the first terrestrial ecosystems is by Jane Gray and William Shear, "Early Life on Land," *American Scientist* 80 (1992): 444–456.

32. N. Campbell, J. Reece, and L. Mitchell, *Biology*, 5th ed. (Menlo Park, Calif.: Benjamin Cummings, 1999).

33. X. Xu et al., "Four-Winged Dinosaurs from China," *Nature* 421 (2003): 335–340.

34. W. H. Beebe, "A Tetrapteryx Stage in the Ancestry of Birds," *Zoologica* 2 (1915): 39–52.

35. X. Xu, Z. Tang, and X. -L. Wang, "A Therizinosauroid Dinosaur with Integumentary Structures from China," *Nature* 399 (1999): 350–354.

36. Mark Pilkington, "Junk DNA: What's in a Name?" *The Guardian*, 22 January 2004.

37. In the spirit of Aristotle, who in his *Politics* said, "Nature does nothing uselessly."

38. J. A. Martens et al., "Intergenic Transcription Is Required to Repress the *Saccharomyces cerevisiae* SER3 Gene," *Nature* 429 (2004): 571–574.

CHAPTER 6

1. M. Rudwick, *Scenes from Deep Time* (Chicago: University of Chicago Press, 1992), 4–16.

2. S. Schama, *Landscape and Memory* (New York: Knopf, 1995), 451.

3. Stephen Jay Gould, *Time's Arrow, Time's Cycle* (Cambridge, Mass.: Harvard University Press, 1987), 30–41.

4. C. Lyell, *Principles of Geology* (London: John Murray, 1830), 37–38.

5. Interestingly, Locke betrayed a deep ambivalence about nature. On the one hand, he described a paradisiacal state of nature in which humans coexisted in an environment of mutual respect and goodwill; on the other hand, nature itself was depicted as stingy and unforgiving. John Locke, *Second Treatise of Government*, ed. Richard Cox (Wheeling, Ill.: Harlan Davidson, 1982; first published in 1689), 27–28.

6. Ibid., 21.

7. Ibid., 24.

8. The complete title of the original edition conveys the ambitiousness of the work: *Manual of Mineralogy, Including Observations on Mines, Rocks, Reduction of Ores, and the Applications of the Science to the Arts.* Dana's *Manual of Mineralogy* is now in its 21st edition, published by Wiley and Sons (1999).

9. R. Laudan, *From Mineralogy to Geology: The Foundations of a Science, 1650–1830* (Chicago: University of Chicago Press, 1987), ch. 3 and 5.

10. James Hutton, *Theory of the Earth,* in *Philosophical Transactions of the Royal Society of Edinburgh,* vol. 1, pt. 2 (1788): 286 and 304.

11. J. W. Powell, *Seventh Annual Report of the Geological Survey to the Two Houses of Congress* (Washington, D.C.: Government Printing Office, 1888), 3–4.

12. See, for example, R. Satz, *Chippewa Treaty Rights* (Madison: Wisconsin Academy of Sciences, Arts and Letters, 1991).

13. D. Abrams, "The Mechanical and the Organic: On the Impact of Metaphor in Science," in *Scientists on Gaia,* ed. S. Schneider and P. Boston (Cambridge, Mass.: MIT Press, 1992), 66–74; Gould, *Time's Arrow, Time's Cycle,* 63–66.

14. Lyell, *Principles of Geology,* 470.

15. Quoted in C. Gillispie, *Pierre-Simon Laplace: A Life in Exact Science* (Princeton, N.J.: Princeton University Press, 1998), 26–27.

16. J. D. Burchfield, *Lord Kelvin and the Age of the Earth* (Chicago: University of Chicago Press, 1990), 13–14.

17. Quoted in P. Appleman, introduction to *Norton Critical Edition of an Essay on the Principle of Population* (New York: Norton, 1976), xi–xxvii.

18. Ibid., xii.

19. For an authoritative description of the influence of Malthus on Darwin's thinking, see Ernst Mayr, *One Long Argument: Charles Darwin and the Genesis of Modern Evolutionary Thought* (Cambridge, Mass.: Harvard University Press, 1991), 84–85.

20. Quoted in M. Kurlansky, *Cod: A Biography of the Fish That Changed the World* (New York: Walker, 1997), 122.

21. Aldo Leopold, *A Sand County Almanac* (Oxford: Oxford University Press, 1948), 214, 204.

22. R. Utgard and G. McKenzie, *Man's Finite Earth* (Minneapolis: Burgess, 1974); C. F. Park, *Earthbound: Minerals, Energy and Man's Future* (San Francisco: Freeman, Cooper and Co., 1975); K. Young, *Geology: The Paradox of*

Earth and Man (Boston: Houghton-Mifflin, 1975); P. Ehrlich, *The Population Bomb* (New York: Ballantine Books, 1968; revised 1978); D. H. Meadows et al., *The Limits to Growth* (New York: Universe Books, 1972); A. Toffler, *Future Shock* (New York: Random House, 1970).

23. Preface to Young, *Geology: The Paradox of Earth and Man.*

24. Quoted in Park, *Earthbound*, 2.

25. J. Tierney, "Betting the Planet," *New York Times Magazine*, 2 December 1990, 52 ff.

26. H. Cole et al., *Models of Doom: A Critique of "The Limits to Growth"* (New York: Universe Books, 1973), ch. 2.

27. Benoit Mandelbrot, "How Long Is the Coastline of Britain? Statistical Self-Similarity and Fractional Dimension," *Science* 156 (1967): 636–638; Benoit Mandelbrot, *The Fractal Geometry of Nature* (San Francisco: W. H. Freeman, 1983).

28. Quoted in A. Moffatt, "Ecologists Look at the Big Picture," *Science* 273 (1996): 1490.

29. Robert Costanza et al., "The Value of the World's Ecosystem Services and Natural Capital," *Nature* 387 (1997): 253–260.

30. H. French, "Investing in the Future: Harnessing Private Capital Flows for Environmentally Sustainable Development," *Worldwatch Paper* 139 (Washington, D.C.: Worldwatch Institute, 1998), 8.

31. E. O. Wilson, *Consilience: The Unity of Knowledge* (New York: Knopf, 1998).

32. Evelyn Fox Keller has pointed out the interesting anachronism inherent in the idea of laws of nature: The metaphor survives from the time when nature was governed by God's edicts. Evelyn Fox Keller, interview by Bill Moyers, *Science and Gender* (Alexandria, Va.: Public Broadcasting System, 1990, videocassette).

33. Chaotic systems are characterized by extreme sensitivity to starting conditions: Trivial differences in the values of variables at the outset can lead to widely diverging results. Nonlinear systems are those in which cause and effect are not in simple linear proportion to each other. Noting that nonlinearity is actually the norm rather than the exception in natural systems, some scientists liken the linear-versus-nonlinear distinction to categorizing animals as elephants and non-elephants.

34. Niles Eldredge, Stephen Jay Gould, J. Coyne, and B. Charlesworth, "On Punctuated Equilibria," *Science* 276 (1997): 338–341.

35. See, for example, W. C. Oechel and G. L. Vourlitis, "The Effects of Climate Change on Arctic Tundra Ecosystems," *Trends in Ecology and Evolution* 9 (1994): 324–329.

36. P. Rabinow, "Chaos in the Garden," *New York Times Book Review*, 12 November 1995.

37. J. Wu and O. L. Loucks, "From Balance of Nature to Hierarchical Patch Dynamics: A Paradigm Shift in Ecology," *Quarterly Review of Biology* 70 (1995): 439–466.

38. James Lovelock, *The Ages of Gaia: A Biography of Our Living Earth* (New York: Norton, 1988).

39. L. Kump, "The Physiology of the Planet," *Nature* 381 (1996): 111–112.

Epilogue

1. Michael Benton, *When Life Nearly Died: The Greatest Mass Extinction of All Time* (New York: Thames and Hudson, 2003), 284–290.

2. C. Elvidge et al., "U.S. Constructed Area Approaches Size of Ohio," *Eos: Transactions, American Geophysical Union* 85 (2004): 233.

3. Concrete production involves heating limestone, which is made of the mineral calcite ($CaCO_3$), to drive off the carbon dioxide bound up in the rock and to make lime (CaO). Although fossil fuel combustion is by far the largest anthropogenic source of carbon dioxide, concrete manufacturing, which returns long-sequestered carbon dioxide to the atmosphere, is also a significant contributor.

INDEX